Engineering
Finite Element Analysis

Synthesis Lectures on Mechanical Engineering

Synthesis Lectures on Mechanical Engineering series publishes 60–150 page publications pertaining to this diverse discipline of mechanical engineering. The series presents Lectures written for an audience of researchers, industry engineers, undergraduate and graduate students. Additional Synthesis series will be developed covering key areas within mechanical engineering.

Engineering Finite Element Analysis
Ramana M. Pidaparti
2017

MEMS Barometers Towards Vertical Position Detection: Background Theory, System Prototyping, and Measurement Analysis
Dimosthenis E. Bolankis
2017

Vehicle Suspension System Technology and Design
Avesta Goodarzi and Amir Khajepour
2017

Engineering Finite Element Analysis

Ramana M. Pidaparti

ISBN: 978-3-031-79569-5   paperback
ISBN: 978-3-031-79570-1   ebook

DOI 10.1007/978-3-031-79570-1

A Publication in the Springer series
SYNTHESIS LECTURES ON MECHANICAL ENGINEERING

Lecture #1
Series ISSN
ISSN pending.

Engineering
Finite Element Analysis

Ramana M. Pidaparti
University of Georgia

SYNTHESIS LECTURES ON MECHANICAL ENGINEERING #1

ABSTRACT

Finite element analysis is a basic foundational topic that all engineering majors need to understand in order for them to be productive engineering analysts for a variety of industries. This book provides an introductory treatment of finite element analysis with an overview of the various fundamental concepts and applications. It introduces the basic concepts of the finite element method and examples of analysis using systematic methodologies based on ANSYS software. Finite element concepts involving one-dimensional problems are discussed in detail so the reader can thoroughly comprehend the concepts and progressively build upon those problems to aid in analyzing two-dimensional and three-dimensional problems. Moreover, the analysis processes are listed step-by-step for easy implementation, and an overview of two-dimensional and three-dimensional concepts and problems is also provided. In addition, multiphysics problems involving coupled analysis examples are presented to further illustrate the broad applicability of the finite element method for a variety of engineering disciplines.

The book is primarily targeted toward undergraduate students majoring in civil, biomedical, mechanical, electrical, and aerospace engineering and any other fields involving aspects of engineering analysis.

KEYWORDS

engineering, matrices, linear algebra, system of equations, modeling, analysis, bar, trusses, beams, vibration, fluid mechanics, solid mechanics, heat transfer, multiphysics, checking, validation and verification

Dedicated to my parents, grandparents, and teachers

Contents

Preface

Engineering analysis based on finite element method concepts is fundamental to many industries due to the complexity of products and problem solving. There is a growing need for fundamental textbooks that integrate the technology/software to better prepare undergraduate engineers for the global industries. This comprehensive book deals with the fundamentals of the finite element method, and the use of ANSYS software in engineering analysis. *Engineering Finite Element Analysis* is appropriate for junior/senior students, and is based on the author's experiences teaching design analysis courses over the past 25 years.

The book is compiled into nine chapters and provides an introduction to basic concepts along with examples to fully illustrate the design analysis and its application to stress, heat transfer, fluid flow, and vibrations. Each of the chapters provides enough detail for students to understand the basics and steps of the finite element analysis process. Several examples are provided that will help students to understand and apply concepts related to design analysis. Chapter 1 provides an introduction to numerical methods, especially finite element analysis, as well as an introduction to ANSYS, a general purpose finite element software. Chapter 2 provides some mathematical background related to matrix algebra, solutions to linear systems of equations, and numerical integration. This chapter discusses the mathematical methods that will be used in later chapters to help students better understand the fundamentals of the finite element method. Chapter 3 describes the finite element basics including formulations, generic finite element procedures, and modeling techniques. Chapter 4 describes the finite element methodology for 1D problems, and provides a comprehensive treatment of problems related to stress, heat transfer, fluid flow, and vibrations, with examples. Chapter 5 discusses the finite element analysis of trusses, beams, and frames, with examples. Chapter 6 provides the analysis of trusses, beams, and frames, with ANSYS software in detail, using screenshots and examples.

Chapter 7 deals with the analysis of 2D problems and provides the elements and shape functions that can be used to discretize a plane 2D geometry. A brief discussion on higher-order elements and axisymmetric elements is also presented. Example problems related to stress, heat transfer, and fluid flow with ANSYS are also presented in Chapter 7. A brief discussion on the 3D finite element analysis with examples using ANSYS is presented in Chapter 8. In addition, an example of a fluid-solid interaction multiphysics problem is presented to further illustrate the analysis. Chapter 9 presents a sample project to better illustrate the finite element process. Several examples of interesting and challenging analysis projects are also presented.

The author acknowledges the support and guidance of many of his colleagues at IUPUI, VCU, and UGA, and extends sincere thanks to many of his students for their help and feedback.

Special thanks to Ed Gaddis, Kittisak Koombua, Israr Ibrahim, JongWon Kim, and Evan Neblett for help with ANSYS examples. Thanks also to Jae-Hwan Lee, Chuck Cartin, John Swanson, Trenicka Rolle, and many other students for their help and contributions. Finally, the author thanks his family (Chitra, Rohan, and Reena) and friends for their support and encouragement.

Ramana M. Pidaparti
May 2017

CHAPTER 1

Introduction

After reading this chapter, you will be able to:

- define Computer-Aided-Engineering (CAE);

- discuss different types of numerical methods;

- explain Finite Element Analysis (FEA);

- identify the importance and benefits of finite element analysis; and

- discuss ANSYS software modules/capabilities.

1.1 OVERVIEW

This chapter provides an introduction to computer-aided-engineering (CAE) analysis using the finite element method. Computer-Aided-Design (CAD) analysis examples as well as salient features of the finite element method are presented to illustrate the design analysis process. An introduction to ANSYS software and its capabilities and major features is provided. ANSYS software graphical user interface is also briefly described.

1.2 COMPUTER-AIDED ENGINEERING

Computers are commonly used in engineering for simulating optimal prototype designs that best meet established performance criteria. Several mathematical and computer science-based tools are frequently used in the construction, evaluation, and testing of design objectives and specifications. It is anticipated that in the very near future, computer modeling methods and tools will merely require a formal description of the desired behavior or structure to enable automatic simulation for product evaluation.

CAE is a technology that uses computers to analyze CAD geometry, which allows the designer to simulate and study how the product will function and behave so that the design can be refined and optimized. CAE software can help perform some of the steps in the design process, especially steps related to geometric modeling, analysis and synthesis. CAE tools/systems are available for a wide range of analyses. These include dynamics analysis, FEA, general purpose, and others. The dynamics analysis includes the kinematics of bodies and deals with motion and force concepts. Several CAE analysis packages such as ANSYS, ABAQUS, COMSOL, and others can calculate the resultant stresses of a design assembly with complex geometry by specifying the loads and using fundamental equations of statics and numerical methods. These packages can be used for design analysis of heat transfer, fluid flow, electromagnetics, piezoelectric, and multi-physics problems. Then, the structural deformation/stresses can be displayed using simulation models. Figure 1.1 shows the overall procedure for CAE and outlines the analysis results used to redesign the part. Design analysis techniques using various types of numerical methods are briefly described in the next section.

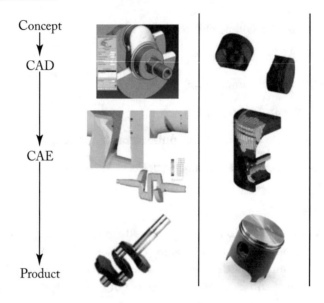

Figure 1.1: CAE approach to product design analysis and synthesis (source: http://www.lozik.h1.ru/).

1.3   TYPES OF NUMERICAL METHODS

Many practical engineering problems involve complex geometry, multiple materials, complex boundary and initial conditions. In order to find a solution to complex engineering problems, we need to resort to numerical methods as it is very difficult to find a closed-form/analytical solution. In general, any physical problem can be idealized as an engineering problem, as shown in Fig. 1.2. Engineering problems are mathematical models of physical situations. Many mathematical models for engineering problems are differential equations, which are derived using fundamental laws of physics (Newton's laws, Fourier law, etc.) and principles (conservation of mass, momentum and energy), with a set of boundary and/or initial conditions. For simple engineering problems, we can find an exact solution in an analytical form where the solution (response) is given in terms of several variables and parameters, as taught in "Mechanics of Materials" or "Elementary Fluid Mechanics" courses. However, for complex problems, numerical methods are required.

These include the Finite Difference Method (FDM), where derivatives are replaced by difference equations and the Finite Element Method (FEM) and the Boundary Element method (BEM) which use integral formulations to create a system of algebraic equations. In FEM, the whole domain/geometry is discretized whereas in the BEM, the boundary of the domain/geometry is discretized. The finite element design analysis techniques are discussed in the next section.

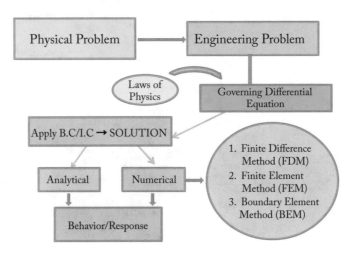

Figure 1.2: An overview of solutions to a physical problem.

1.4   FINITE ELEMENT ANALYSIS

Finite element analysis (FEA) is a general numerical method capable of solving a variety of engineering problems and the underlying theory is more than 100 years old, dating back to early 1900s. Major companies including Boeing, Ford, General Electric, Intel, IBM, Apple, and others employ general purpose computer software to efficiently complete daily engineering analysis tasks.

Finite element analysis is a numerical method that predicts the behavior (response) of a product subjected to loads (inputs). FEA is very popular and can find applications in the design analysis of mechanical, aerospace, biomedical, civil, and electrical systems. In FEA, the geometry of the designed part is broken down into smaller elements that are interconnected at nodes. In discretizing the part, the entire product geometry is filled with elements without any overlaps, and analyzed for functional performance. The part (example, an aircraft wing) is divided into a finite element mesh with smaller elements connected at nodes, as shown in Fig. 1.3. After applying loads and boundary conditions to the part, the resulting finite element equations are solved. FEA can be used for many types of analysis including stress, heat transfer, fluid flow, buckling, vibrations and multi-physics involving more than one discipline. The factor of safety against failure can be predicted from the stress analysis. Fig. 1.4 provides several design simulation examples to illustrate the applicability of FEA for analyzing a variety of engineering problems.

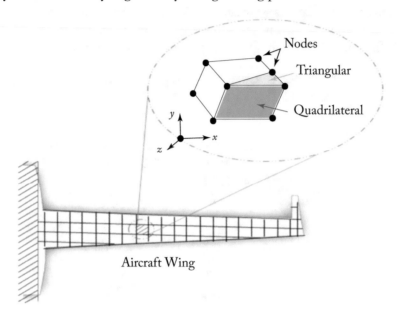

Figure 1.3: An aircraft wing is discretized with finite elements defined by shape/nodes.

1.4.1   THE NEED FOR FEA

FEA is a numerical method of solving systems of equations after discretizing the geometry with finite elements and finding a solution after applying boundary conditions to the assembled system. After formulating the finite element analysis problem, the resulting equations can be arranged for a linear system as

$$[A]\{x\} = \{b\},$$

where $[A]$ is the system matrix, $\{x\}$ is the response and $\{b\}$ is the input to the system. In stress analysis, the matrix $[A]$ becomes the stiffness matrix, vector $\{b\}$ becomes the force vector, and vector $\{x\}$ becomes the displacement vector. The method for solving response $\{x\}$ is discussed in detail in Chapter 2. In summary, the following are the main reasons for conducting design analysis using the finite element method:

- to reduce the amount of prototype testing;

- computer simulation allows multiple "what-if" scenarios to be tested quickly and effectively; and

- to simulate designs (surgical implants, Mars rover, oil rigs, etc.) that are not suitable for prototype testing or experimentation.

The bottom line:

- cost savings in the design cycle;

- time savings…reduce time to market!; and

- create more reliable and better-quality designs.

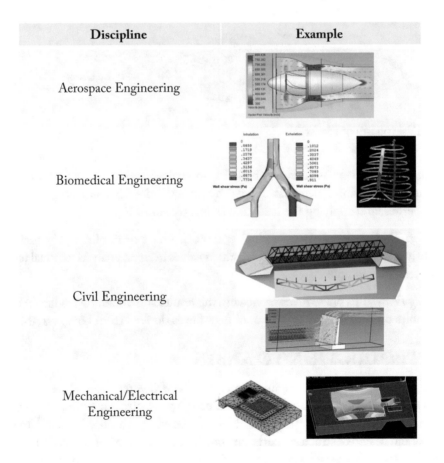

Discipline	Example
Aerospace Engineering	
Biomedical Engineering	
Civil Engineering	
Mechanical/Electrical Engineering	

Figure 1.4: Examples of FEA in many engineering disciplines.

1.4.2 BASIC STEPS IN FEA

The objective of FEA is to determine the unknowns (degrees of freedom) at the nodes and the resulting support reactions in any structure. There are three basic steps or phases involved in finite element analysis, as shown in Fig. 1.5.

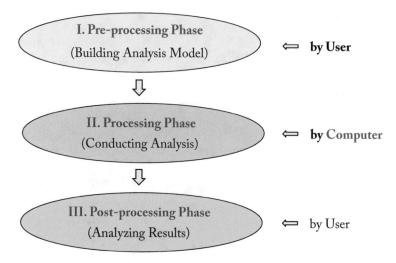

Figure 1.5: Basic phases of the FEM.

Pre-processing Phase: This phase involves building an analysis model by discretizing the domain geometry into specific finite elements defined by nodes and their connections, specifying the material properties, and applying the loads and boundary conditions.

Solution Phase: This phase involves developing a set of linear or nonlinear algebraic equations simultaneously to obtain nodal results (displacement values in stress analysis or nodal temperatures in heat transfer), for example.

Post-processing Phase: This phase involves viewing the results and obtaining results on other quantities or variables of interest such as stresses or heat fluxes derived from nodal variables.

1.5 INTRODUCTION TO ANSYS

ANSYS is a general-purpose finite element software that is tailored for analysis of multi-physics problems in solid mechanics, heat transfer, fluid mechanics, electromagnetics, and acoustics. It is a commercially available computer program which may be extensively used for studying the structural integrity and design of structural parts, and efficiency and design of thermal, fluid, magnetic, and electrical systems. It is used by mechanical, civil, aerospace, electrical, and chemical engineers in various industries. More specifically, ANSYS enables engineers to perform the following tasks:

- Build computer models or transfer CAD models of structures, products, components, or systems.

- Apply operating loads or design performance conditions.

- Study the physical responses, such as stress levels, temperature distributions, or the impact of electromagnetic fields.

- Optimize a design early in the development process to reduce production costs.

- Perform prototype testing in environments where it otherwise would be undesirable or impossible (for example biomedical applications).

The ANSYS software has a comprehensive graphical user interface that gives users easy, interactive access to program functions, commands, documentation, and reference material.

1.5.1   ANSYS MODULES

ANSYS software has the following modules.

ANSYS/Multiphysics. This is a most comprehensive product which includes all other ANSYS modules. Capabilities include structural, thermal, fluid, electromagnetic, and acoustic analyses. Coupling between structural, acoustic, electromagnetic, fluid, and thermal fields is also possible.

ANSYS/Structural. This module includes structural analysis capabilities for static and dynamic stress analysis of structures including the buckling and geometrical/material nonlinearities encountered in certain structures.

ANSYS/Mechanical. This module includes ANSYS/Structural with additional capabilities for linear/nonlinear and steady/transient heat transfer and acoustics analysis capabilities. Coupled thermal-structural and acoustic-structural analyses can also be performed.

ANSYS/LinearPlus. This module is restricted to linear static and dynamic analysis of structures.

ANSYS/Thermal. This module is restricted to steady/transient and linear/nonlinear heat transfer analysis of solids.

ANSYS/Emag. This module is designed for electromagnetics and electrostatics analysis of, electrical systems, motors, alternators, radars, etc.

ANSYS/Flotran. This Computational Fluid Dynamics Module (CFD) analyzes viscous, turbulent, incompressible and compressible fluid flow problems with and without heat transfer, including models for multi-species.

1.5.2   USING ANSYS

The simplest way to enter the ANSYS program is through the ANSYS Launcher, as shown in Fig. 1.6. The launcher has a menu containing push buttons that provide the choices you need to run the ANSYS program and other auxiliary programs.

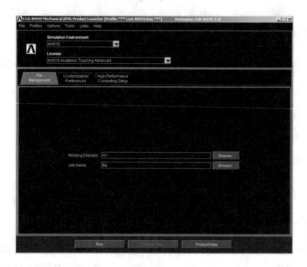

Figure 1.6: The ANSYS Launcher dialog box.

When using the Launcher to enter ANSYS, follow these basic steps:

- From the Start menu select **Programs → ANSYS 14.0 → Mechanical APDL Product Launcher**.

- The ANSYS Launcher dialog box will appear, containing interactive entry options.

Working directory.   This directory is the one in which the ANSYS run will be executed. If the directory displayed is not the one you want to work in pick the "**Browse...**" button to the right of the directory name and specify the desired directory. Note the directories you can use on engineering computers are on your home directory, **H:**, and USB flash drive **E:**.

Jobname.   The jobname is the one that will be used as the prefix of the file name for all files generated by the ANSYS run. Type the desired jobname in this field of the dialog box (the extension for the file names will be .DB).

GUI configuration.   This command brings up a dialog box that allows you to choose the desired menu layout and font size. Do not change the default settings, but simply press OK on this dialog box so that the proper settings file is created for the terminal you are using. This step is only required the first time you enter ANSYS.

A typical analysis in ANSYS involves three distinct steps.

1. *Preprocessing*. Using the **Preprocessor**, you provide data such as the geometry, materials, and element type.

   - Specify job name and title.
   - Set preferences.
   - Define element types and options.
   - Define real constants.
   - Define material properties.
   - Create model geometry.
   - Define mesh the entities.

2. *Solution*. Using the **Solution** processor, you define the type of analysis, set boundary conditions, apply loads, and initiate finite element solutions.

   - Apply Boundary Condition.
   - Specify solution controls (Static /Dynamic, Heat Transfer, etc.).
   - Specify transient characteristics—More to come later.
   - Solve.

3. *Postprocessing*. Using **General Postprocessor** (for static or steady state problems) or **Time-Hist Postprocessor** (for transient problems), you review the results of your analysis through graphical displays and tabular listings.

   You enter a processor by selecting it from the **Preferences** submenu of ANSYS **Main** menu in the GUI. You can move from one processor to another by simply choosing the processor you want from the ANSYS main menu.

1.5.3   GRAPHICAL USER INTERFACE (GUI)

The simplest way to communicate with ANSYS is by using the ANSYS menu system, the GUI, as shown in Fig. 1.7. The GUI provides an interface between you and the ANSYS program. The program is internally driven by ANSYS commands. However, by using the GUI, you can perform an analysis with little or no knowledge of ANSYS commands. This process works because each GUI function ultimately produces one or more ANSYS commands that are automatically executed by the program.

Layout of the GUI.   The ANSYS GUI consists of seven regions, as shown in Fig. 1.7. They include the following.

- **Utility Menu**—Contains utility functions that are available throughout the ANSYS session, such as file controls, selecting, graphics controls, and parameters. You also exit the ANSYS program through this menu.

- **Command Input Area**—Allows you to type in commands directly. All previously typed in commands can be viewed using the drop down arrow to the right of the command input area.

- **Main Menu**—Contains the primary ANSYS functions, organized by processors (preprocessor, solution, general postprocessor, design optimizer, etc.), as shown in Fig. 1.8.

- **Output Window**—Receives text output from the program. It is usually positioned behind the other windows, but you can bring it to the front when necessary.

- **Toolbar**—Contains push buttons that execute commonly used ANSYS commands and functions. You may add your own push buttons by defining abbreviations.

- **Graphics Window**—A window where graphics displays are drawn.

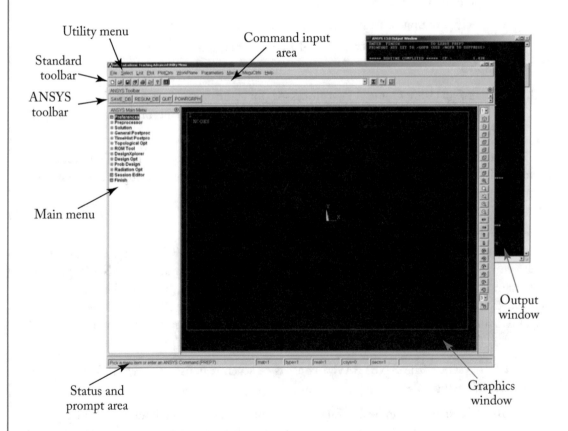

Figure 1.7: The ANSYS-GUI.

Figure 1.8: ANSYS main menu.

- **Status and Prompt Area** —Shows the status of the current analysis and prompts user for appropriate input.

1.5.4   ANSYS WORKBENCH

The recent versions of ANSYS software uses "Workbench" in addition to "Mechanical." ANSYS Workbench streamlines different components (pre- and post-processing tools and solvers) of ANSYS physics simulation. It also enables easier user-interfacing for multi-physics and coupled problems. Figure 1.9 shows the user interface of Workbench. The Workbench showcases ANSYS physics simulation pre- and post-processor, and solvers in one place. These are called "components." Users can drag a component on the right to use it, and connect components for multi-physics or coupled simulation.

As can be seen from Fig. 1.9, the components in ANSYS Workbench are categorized according to the problem to be solved. The components provide improved user interface for traditional ANSYS FEA tools such as Mechanical. In addition, models can be built using External CAD software (as well as ANSYS built-in software, the Design Modeler that can be found in "Geometry" component), and imported it to the component. Figure 1.10 shows user interface for "Static Structural" component, that uses ANSYS Mechanical as its backend to solve a static elasticity problem.

1.5.5   ANSYS WORKBENCH VS. ANSYS MECHANICAL

In the latest version of ANSYS software (beyond version 13), the Workbench option is introduced to conduct multi-physics and multi-software analysis. The examples presented in this book are based on "ANSYS Mechanical." The differences between the ANSYS Workbench and ANSYS Mechanical in terms of the process as well as GUI are presented in Figs. 1.11 and 1.12, respectively.

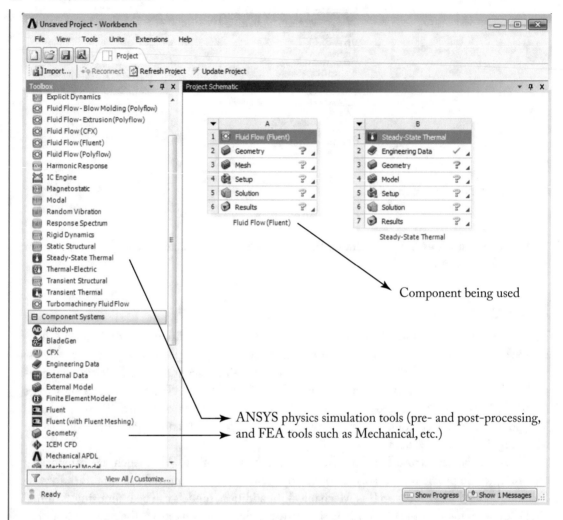

Figure 1.9: ANSYS Workbench user interface consists of components on the right that can be dragged to Workbench field from the left to be used in the solution. The components on the right consists of traditional ANSYS physics simulation solvers (such as mechanical, fluent) with improved user interface.

Note: In the ANSYS program, we can use any system of units for the problem as long as they are consistent (either SI or British system), as the program does not assume any system of units.

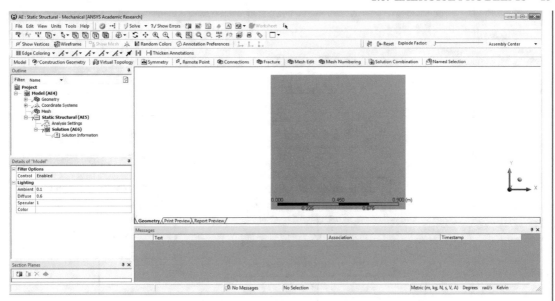

Figure 1.10: User interface for static structural analysis in Workbench.

1.6   EXERCISE PROBLEMS

1.1.   What is CAE and why should we use it?

1.2.   Identify five design analysis problems from everyday life.

1.3.   What are the major steps/phases in FEA?

1.4.   Select a product and determine the design analysis objectives. Discuss how the design analysis can be performed to improve product function.

1.5.   Find two examples of designs and describe the design analysis steps.

1.6.   Find design analysis examples related to civil engineering, mechanical engineering, and aerospace engineering.

1.7.   What is ANSYS software? Discuss its capabilities for engineering design.

1.8.   What are some of the major advantages in employing ANSYS software in a company for product design?

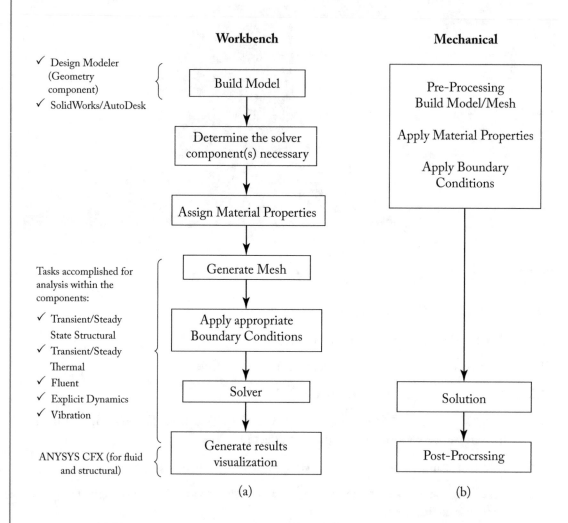

Figure 1.11: (a) Typical work flow of FEA using ANSYS Workbench. The notes show the components or external software (if necessary) to use for each step. Some components, such as Structural and Thermal, use ANSYS Mechanical as solver. (b) Typical work flow of FEA using ANSYS Mechanical. Each step is accomplished within ANSYS Mechanical.

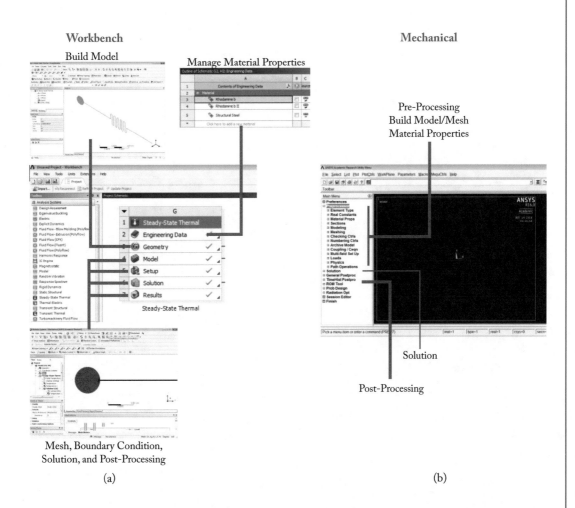

Figure 1.12: (a) Work flows in **ANSYS Workbench** are based on components that are modular. Typical components consist of model builder, material data manager, and user interface for meshing, solver, and post-processor. The modular design allows for easy data interchanges between component/solver. (b) Works in **ANSYS Mechanical** are confined within the user interface.

CHAPTER 2

Mathematical Preliminaries

After reading this chapter, you will be able to:

- explain matrices and their basic operations;

- find a solution to a matrix system of equations;

- find eigenvalues and eigenvectors of a matrix/system;

- evaluate an integral numerically; and

- use MATLAB to solve matrix algebra problems.

2.1   OVERVIEW

This chapter introduces matrix algebra, and discusses example procedures for solving matrix systems of equations and eigenvalue problems. An introduction to numerical integration is also presented as it is commonly used in finite element analysis. MATLAB instructions for carrying out matrix operations are also presented.

2.2   MATRIX ALGEBRA

FEA uses matrix algebra and vector calculus extensively, therefore an introduction to matrix algebra is presented in this chapter. Also, a matrix is a convenient way to represent engineering data. Here, we review a set of matrix operations and functions that apply to finite element analysis.

Row Vector: The row vector is a collection of elements (scalars) arranged in a row that has many columns (n). See for example, row vector $\{A\}$ as shown below:

$$\{A\} = \text{elements that are arranged } (1xn) \text{ as,}$$
$$\{A\}_{1xn} = \{a_1, \ a_2, \ \dots \dots .a_n\}.$$

Column Vector: The column vector is a collection of elements (scalars) arranged in a column that has many rows (m).

$$\{B\} = \text{elements that are arranged } (mx1) \text{ as,} \qquad \{B\}_{mx1} = \left\{ \begin{array}{c} b_1 \\ b_2 \\ \dots \\ \dots \\ .. \\ b_m \end{array} \right\}.$$

Matrix: A matrix consists of an array of mathematical objects arranged in rows and columns. The matrix size is defined by its rows and columns. Thus, a matrix of size $m \times n$ has m rows and n columns. Matrix $[A]$ with 3 rows and 3 columns with elements is shown below.

$$A = \left[\begin{array}{ccc} a_{11} & a_{12} & a_{13} \\ a_{21} & a_{22} & a_{23} \\ a_{31} & a_{32} & a_{33} \end{array} \right].$$

2.2.1   MATRIX ADDITION AND SUBTRACTION

If matrix A and matrix B are $m \times n$ matrices, then the matrix addition or subtraction gives

$$C = A \pm B,$$

where C is also a $m \times n$ matrix (compatibility condition—numbers of columns of A and number of rows of B should be the same). This can be accomplished by adding or subtracting element by element as,

$$C_{ij} = A_{ij} \pm B_{ij} \text{ for } i = 1, 2, \dots m \text{ and } j = 1, 2, \dots n.$$

Example: For $m = 2$ and $n = 2$,

$$\overset{[A]}{\left[\begin{array}{cc} a_{11} & a_{12} \\ a_{21} & a_{22} \end{array} \right]} + \overset{[B]}{\left[\begin{array}{cc} b_{11} & b_{12} \\ b_{21} & b_{22} \end{array} \right]} = \left[\begin{array}{cc} a_{11} + b_{11} & a_{12} + b_{12} \\ a_{21} + b_{21} & a_{22} + b_{22} \end{array} \right]$$

$$\overset{[A]}{\left[\begin{array}{cc} a_{11} & a_{12} \\ a_{21} & a_{22} \end{array} \right]} - \overset{[B]}{\left[\begin{array}{cc} b_{11} & b_{12} \\ b_{21} & b_{22} \end{array} \right]} = \left[\begin{array}{cc} a_{11} - b_{11} & a_{12} - b_{12} \\ a_{21} - b_{21} & a_{22} - b_{22} \end{array} \right]$$

Note: multiplication of matrix $[A]$ and matrix $[B]$ is not the same as the multiplication of matrix $[B]$ and matrix $[A]$. That is,

$$[A][B] \neq [B][A].$$

2.2.2   SCALAR MULTIPLICATION

When a scalar quantity "α" is multiplied with a matrix $[A]$, the resulting matrix is the same size as $[A]$ but the elements are the product of elements α times the elements of $[A]$ as given below:

$$\alpha * [A]_{m \times n} = \alpha * a_{ij}.$$

Example:

$$5 \begin{bmatrix} 2 & 0 \\ 1 & 2 \end{bmatrix} = \begin{bmatrix} 10 & 0 \\ 5 & 10 \end{bmatrix}.$$

2.2.3   TRANSPOSE OF A MATRIX

The transpose of a matrix $[A]$ is obtained by interchanging its rows and columns and is denoted by

$$B = A^T \text{ or } B_{ij} = A_{ji}.$$

Example:

$$A = \begin{bmatrix} a_{11} & a_{12} \\ a_{21} & a_{22} \\ a_{31} & a_{32} \end{bmatrix}, \quad A^T = \begin{bmatrix} a_{11} & a_{21} & a_{31} \\ a_{12} & a_{22} & a_{32} \end{bmatrix}.$$

Square Matrix: A matrix is defined as a square matrix if the number of rows is equal to the number of columns.

$$[A]_{n \times n} \rightarrow \begin{bmatrix} 2 & 1 \\ 1 & 2 \end{bmatrix}_{2 \times 2} \rightarrow \begin{bmatrix} 3 & 1 & 1 \\ 2 & 2 & 3 \\ 4 & 1 & 4 \end{bmatrix}_{3 \times 3}.$$

Diagonal Matrix: A matrix $[A]$ is defined as a diagonal matrix if all the elements of a square matrix are equal to zero, except those on the principal diagonal.

$$[A] = \begin{bmatrix} a_{11} & 0 & 0 & 0 & 0 \\ 0 & a_{22} & 0 & 0 & 0 \\ 0 & 0 & a_{33} & 0 & 0 \\ 0 & 0 & 0 & a_{44} & 0 \\ 0 & 0 & 0 & 0 & a_{nn} \end{bmatrix}_{n \times n}.$$

Unit or Identity Matrix: A unit matrix is a special case of a diagonal matrix where all the diagonal elements are equal to 1 and all off-diagonal elements are equal to zero.

Symmetric Matrix: A symmetric matrix is defined as a square matrix with property, for i not equal to j, as

$$a_{ij} = a_{ji}.$$

2.2.4   MULTIPLICATION OF VECTORS

Let "a" and "b" represent two vectors of size m (or $m \times 1$), then the *Dot* or *Inner Product* is defined as

$$c = a^T b = b^T a = \sum_{i=1}^{m} (a_i b_i)$$
$$= a_1 b_1 + a_2 b_2 + \cdots + a_n b_n.$$

The *Tensor Product* is defined as

$$D = a b^T$$

or

$$D_{ij} = a_i b_j \text{ for } i, j = 1, 2 \ldots m.$$

Example: for $m = 3$,

$$D = \begin{bmatrix} a_1 b_1 & a_1 b_2 & a_1 b_3 \\ a_2 b_1 & a_2 b_2 & a_2 b_3 \\ a_3 b_1 & a_3 b_2 & a_3 b_3 \end{bmatrix}.$$

2.2.5   MULTIPLICATION OF MATRICES

If the size of matrix $[A]$ is a $m \times l$ and the size of matrix $[B]$ is a $l \times n$, then the matrix multiplication gives

$$C = A B,$$

where $[C]$ is a $m \times n$ matrix. This can be accomplished by

$$C_{ij} = \sum_{k=1}^{m} a_{ik} b_{kj}$$

for $i = 1, 2 \ldots l$ and $j = 1, 2 \ldots n$.

Example: for $m = 2, l = 3, n = 2$, then

$$C_{i,j} = a_{i1} b_{1j} + a_{i2} b_{2j} + a_{i3} b_{3j}$$
$$C_{11} = a_{11} b_{11} + a_{12} b_{21} + a_{13} b_{31}$$
$$C_{12} = a_{11} b_{12} + a_{12} b_{22} + a_{13} b_{32}$$
$$C_{21} = a_{21} b_{11} + a_{22} b_{21} + a_{23} b_{31}$$
$$C_{21} = a_{21} b_{12} + a_{22} b_{22} + a_{23} b_{32}.$$

2.2.6   MULTIPLICATION OF A MATRIX WITH A VECTOR

If $[A]$ is a $m \times n$ matrix and x is a $n \times 1$ vector, then

$$A x = f,$$

where $\{f\}$ is a $m \times 1$ vector. This can be accomplished by

$$f_i = \sum_{j=1}^{n} A_{ij} x_j$$

for $i = 1, 2 \ldots m$.

2.2.7   DETERMINANT

A **determinant** is a scalar value computed from the elements in a square matrix. Determinants have various applications in engineering, including computing inverses and solving systems of simultaneous equations. For a 2×2 matrix,

$$A = \left[\begin{array}{cc} a_{11} & a_{12} \\ a_{21} & a_{22} \end{array} \right]$$

the determinant is

$$|A| = a_{11}a_{22} - a_{21}a_{12}.$$

For a 3×3 matrix

$$A = \left[\begin{array}{ccc} a_{11} & a_{12} & a_{13} \\ a_{21} & a_{22} & a_{23} \\ a_{31} & a_{32} & a_{33} \end{array} \right]$$

the determinant is:

$$|A| = a_{11}a_{22}a_{33} + a_{12}a_{23}a_{31} + a_{13}a_{21}a_{32}$$
$$- a_{31}a_{22}a_{13} - a_{32}a_{23}a_{11} - a_{33}a_{21}a_{12}.$$

2.2.8   INVERSE OF A MATRIX

If $[A]$ is a square matrix of size m, then the inverse of A is denoted as A^{-1} which satisfies the relationship

$$I = AA^{-1} = A^{-1}A,$$

where I is an identity matrix (square).

For inverse of a matrix to exist, its determinant must be non-zero. Matrices with zero determinants are called *singular matrices*. There are a number of methods for evaluating inverse of matrices.

For example, the inverse of a 2×2 matrix $[A]$ is:

$$A^{-1} = \frac{1}{|A|} \left[\begin{array}{cc} a_{22} & -a_{12} \\ -a_{21} & a_{11} \end{array} \right].$$

2.3   SYSTEM OF LINEAR EQUATIONS

Consider a system of equations arranged as $[A]\{x\} = \{f\}$. The above equations denote a system of n linear equations with n unknowns x. In that case, A_{ij} is called the coefficient matrix of the system of equations.

Example: For $n = 3$,

$$\begin{bmatrix} a_{11} & a_{12} & a_{13} \\ a_{21} & a_{22} & a_{23} \\ a_{31} & a_{32} & a_{33} \end{bmatrix} \begin{bmatrix} x_1 \\ x_2 \\ x_3 \end{bmatrix} = \begin{bmatrix} f_1 \\ f_2 \\ f_3 \end{bmatrix}$$

or

$$a_{11}x_1 + a_{12}x_2 + a_{13}x_3 = f_1$$
$$a_{21}x_1 + a_{22}x_2 + a_{23}x_3 = f_2 \ .$$
$$a_{31}x_1 + a_{32}x_2 + a_{33}x_3 = f_3$$

Note: If the determinant of A is non-zero (i.e., if A is non-singular), the unknown vector x is calculated by pre-multiplying both sides of the system of equations by A^{-1} to obtain

$$A^{-1}Ax = Ix = x = A^{-1}f.$$

In addition to matrix inversion method, there are also a number of other methods available for solving systems of linear equations, such as Cramer's rule, LU decompositions, Gaussian eliminations, etc. These methods are briefly described below.

2.3.1   CRAMER'S RULE

Consider the following equation $[A]\{x\} = \{f\}$ with $n = 3$. The solution to the equation with Cramer's rule can be obtained as follows:

$$x_1 = \frac{\begin{vmatrix} f_1 & a_{12} & a_{13} \\ f_2 & a_{22} & a_{23} \\ f_3 & a_{32} & a_{33} \end{vmatrix}}{|A|}; x_2 = \frac{\begin{vmatrix} a_{11} & f_1 & a_{13} \\ a_{21} & f_2 & a_{23} \\ a_{31} & f_3 & a_{33} \end{vmatrix}}{|A|}; x_3 = \frac{\begin{vmatrix} a_{11} & a_{12} & f_1 \\ a_{21} & a_{22} & f_2 \\ a_{31} & a_{32} & f_3 \end{vmatrix}}{|A|}.$$

2.3.2   GAUSSIAN ELIMINATION

This is an organized method of substituting each equation into a previous equation until the last equation contains only one equation. Then the unknowns are determined sequentially, starting with the last equation and proceeding upward to the first equation. Let us consider the following two equations:

$$a_{11}x_1 + a_{12}x_2 = b_1 \tag{2.1}$$
$$a_{21}x_1 + a_{22}x_2 = b_2 \tag{2.2}$$

$$\underbrace{\frac{EQ(1)}{a_{11}}}_{} - \underbrace{\frac{EQ(2)}{a_{21}}}_{} \qquad x_2\left(\frac{a_{12}}{a_{11}} - \frac{a_{22}}{a_{21}}\right) = \left(\frac{b_1}{a_{11}} - \frac{b_2}{a_{21}}\right).$$

Eliminate x_1

Results in

$$x_2 = \frac{\left(\dfrac{b_1}{a_{11}} - \dfrac{b_2}{a_{21}}\right)}{\left(\dfrac{a_{12}}{a_{11}} - \dfrac{a_{22}}{a_{21}}\right)}.$$

Similarly, we can find,

$$x_1 = \frac{\left(\dfrac{b_1}{a_{12}} - \dfrac{b_2}{a_{22}}\right)}{\left(\dfrac{a_{11}}{a_{12}} - \dfrac{a_{21}}{a_{22}}\right)}.$$

2.3.3   CHOLESKY'S DECOMPOSITION

The Cholesky's decomposition method involves decomposing the given matrix $[A]$ into a lower triangular matrix $[L]$ and upper triangular matrix $[U]$ and formulating the solution. This method is particularly helpful for solving stress analysis problems where the response is required for multiple load combinations (right-hand-side matrix). The procedure is given below.

Given $[A]\{x\} = \{b\}$.

Decompose $[A] = [L] * [U]$, then substituting back into original equation,

$$[L]_{n \times n}[U]_{n \times n}\{x\}_{n \times 1} = \{b\}$$

will result in

$$[L]\{z\} = \{b\}$$
$$\{z\} = [L]^{-1}\{b\}$$
$$\{x\} = [U]^{-1}\{z\}$$
$$\{x\} = [U]^{-1}[L]^{-1}\{b\}.$$

Consider matrix $[A]$ of size 3×3 as given below:

$$A = \begin{bmatrix} a_{11} & a_{12} & a_{13} \\ a_{21} & a_{22} & a_{23} \\ a_{31} & a_{32} & a_{33} \end{bmatrix}.$$

Let us say matrix A is a square matrix that is symmetric and positive definite, then the upper triangular matrix U can be chosen such that $U = L^T$, where L is the lower triangular matrix. Then the elements of L matrix can be calculated as below.

The formula for entries of L is,

$$l_{11} = \sqrt{a_{11}}$$

$$l_{j1} = \frac{a_{j1}}{l_{11}} \qquad j = 2, \ldots, n$$

$$l_{jj} = \sqrt{a_{jj} - \sum_{s=1}^{j-1} l_{js}^2} \qquad j = 2, \ldots, n$$

$$l_{pj} = \frac{1}{l_{jj}} \left(a_{pj} - \sum_{s=1}^{j-1} l_{js} l_{ps} \right) \qquad p = j + 1, \ldots, n; \qquad j \geq 2.$$

Example 2.1   Solve the following set of equations (a) using the Gaussian method and (b) using the LU decomposition method:

$$2x_1 - 3x_2 = 1 \tag{1}$$

$$5x_1 + x_2 = 2. \tag{2}$$

Solution:

Gaussian method:

Divide (1) by 2 and (2) by 5

$$x_1 - \frac{3}{2}x_2 = \frac{1}{2} \tag{1a}$$

$$x_1 + \frac{1}{5}x_2 = \frac{2}{5}. \tag{2a}$$

Subtract (2a) from (1a)

$$\frac{-17}{10}x_2 = \frac{1}{10}$$

$$x_2 = -\frac{1}{17}.$$

From (1)

$$2x_1 + \frac{3}{17} = 1$$

$$2x_1 = \frac{14}{17}$$

$$x_1 = \frac{7}{17}.$$

LU decomposition:

$$[A]\{x\} = \{b\}$$

$$\begin{bmatrix} 2 & -3 \\ 5 & 1 \end{bmatrix} \begin{Bmatrix} x_1 \\ x_2 \end{Bmatrix} = \begin{Bmatrix} 1 \\ 2 \end{Bmatrix}$$

$$[L] = \begin{bmatrix} 1 & 0 \\ l_{21} & 1 \end{bmatrix} \quad [U] = \begin{bmatrix} u_{11} & u_{12} \\ 0 & u_{22} \end{bmatrix}$$

$$u_{11} = a_{11} = 2; \quad u_{12} = a_{12} = -3$$

$$l_{21}u_{11} = a_{21} = 5; \quad l_{21} = \frac{5}{2}$$

$$l_{21}u_{12} + u_{22} = a_{22} = 1$$

$$-\frac{15}{2} + u_{22} = 1; \quad u_{22} = \frac{17}{2}$$

$$[A] = [L][U]$$

$$\begin{bmatrix} 2 & -3 \\ 5 & 1 \end{bmatrix} = \begin{bmatrix} 1 & 0 \\ \frac{5}{2} & 1 \end{bmatrix} \begin{bmatrix} 2 & -3 \\ 0 & \frac{17}{2} \end{bmatrix}$$

$$\{z\} = [L]^{-1}\{b\} = \begin{bmatrix} 1 & 0 \\ \frac{5}{2} & 1 \end{bmatrix}^{-1} \begin{Bmatrix} 1 \\ 2 \end{Bmatrix}$$

$$= \frac{1}{|L|} \begin{bmatrix} 1 & 0 \\ -\frac{5}{2} & 1 \end{bmatrix} \begin{Bmatrix} 1 \\ 2 \end{Bmatrix} = \begin{Bmatrix} 1 \\ -\frac{1}{2} \end{Bmatrix}$$

$$\{x\} = [U]^{-1}\{z\} = \frac{1}{|U|} \begin{bmatrix} \frac{17}{2} & 3 \\ 0 & 2 \end{bmatrix} \begin{Bmatrix} 1 \\ -\frac{1}{2} \end{Bmatrix} = \frac{1}{17} \begin{Bmatrix} 7 \\ -1 \end{Bmatrix}$$

$$\{x\} = \begin{Bmatrix} \frac{7}{17} \\ -\frac{1}{17} \end{Bmatrix}.$$

Example 2.2   Given:

$$A = \begin{bmatrix} 3 & -4 \\ 6 & -10 \end{bmatrix}, \quad x = \begin{Bmatrix} x_1 \\ x_2 \end{Bmatrix}, \quad \{b\} = \begin{Bmatrix} 5 \\ 2 \end{Bmatrix}; \quad \text{find } x = ?$$

Solution:

$$[A] = [L][U]$$

$$\begin{bmatrix} 3 & -4 \\ 6 & -10 \end{bmatrix} = \begin{bmatrix} 1 & 0 \\ L_{21} & 1 \end{bmatrix} \begin{bmatrix} U_{11} & U_{12} \\ 0 & U_{22} \end{bmatrix}$$

$$\textit{Left side} = \textit{right side}$$

$$3 = 1\,(U_{11}) + (0)\,(0) \rightarrow U_{11} = 3$$

$$-4 = 1\,(U_{12}) + (0)\,(U_{22}) \rightarrow U_{12} = -4$$

$$6 = (L_{12})\,(U_{11}) + 1\,(0) \rightarrow L_{12} = \frac{6}{U_{11}} = \frac{6}{3} = 2$$

$$-10 = (L_{21})\,(U_{12}) + 1\,(U_{22}) \rightarrow U_{22} = -2$$

$$[A] = [L]\,[U] \quad \begin{bmatrix} 1 & 0 \\ 2 & 1 \end{bmatrix} \begin{bmatrix} 3 & -4 \\ 0 & -2 \end{bmatrix}$$

$$\{z\} = [L]^{-1}\{b\} = \begin{bmatrix} 1 & 0 \\ 2 & 1 \end{bmatrix}^{-1} \begin{Bmatrix} 5 \\ 2 \end{Bmatrix} = \frac{1}{|L|}\begin{bmatrix} 1 & 0 \\ -2 & 1 \end{bmatrix}\begin{Bmatrix} 5 \\ 2 \end{Bmatrix} = \begin{Bmatrix} 5 \\ -8 \end{Bmatrix}$$

$$\{x\} = [U]^{-1}\{z\} = \frac{1}{|U|}\begin{bmatrix} 3 & -4 \\ 0 & -2 \end{bmatrix}\begin{Bmatrix} 5 \\ 2 \end{Bmatrix} = \begin{Bmatrix} 7 \\ 4 \end{Bmatrix}$$

Example 2.3   Solve the following system of equations using Cholesky's decomposition method.

$$4x_1 + 6x_2 + 8x_3 = 0$$
$$6x_1 + 34x_2 + 52x_3 = -160$$
$$8x_1 + 52x_2 + 129x_3 = -452.$$

Write the equations in matrix form:

$$\begin{bmatrix} 4 & 6 & 8 \\ 6 & 34 & 52 \\ 8 & 52 & 129 \end{bmatrix} \begin{bmatrix} x_1 \\ x_2 \\ x_3 \end{bmatrix} = \begin{bmatrix} 0 \\ -160 \\ -452 \end{bmatrix}.$$

Calculate the entries of L matrix using the above equations:

$$l_{11} = \sqrt{4} = 2$$

$$l_{j1} = \frac{a_{j1}}{l_{11}} \quad \text{for} \quad j = 2, 3$$

$$l_{21} = \frac{a_{21}}{l_{11}} = \frac{6}{2} = 3 \qquad l_{31} = \frac{8}{2} = 4$$

$$l_{jj} = \sqrt{a_{jj} - \sum_{s=1}^{j-1} l_{js}^2} \quad \text{for} \quad j = 2$$

$$l_{22} = \sqrt{a_{22} - \sum_{s=1}^{1} l_{2s}^2} = \sqrt{34 - (3^2)} = 5$$

$$l_{pj} = \frac{1}{l_{jj}} \left(a_{pj} - \sum_{s=1}^{j-1} l_{js} l_{ps} \right) \quad \text{with} \quad p = 3; \quad j \geq 2$$

$$l_{32} = \frac{1}{l_{22}} \left(a_{32} - \sum_{s=1}^{1} l_{2s} l_{3s} \right) = \frac{1}{5} (52 - (3 \times 4)) = 8$$

$$l_{jj} = \sqrt{a_{jj} - \sum_{s=1}^{j-1} l_{js}^2} \quad \text{for} \quad j = 2$$

$$l_{33} = \sqrt{a_{33} - \sum_{s=1}^{2} l_{3s}^2} = \sqrt{129 - (4^2 + 8^2)} = 7.$$

The matrix A becomes

$$L = \begin{bmatrix} 2 & 0 & 0 \\ 3 & 5 & 0 \\ 4 & 8 & 7 \end{bmatrix} \qquad L^T = U = \begin{bmatrix} 2 & 3 & 4 \\ 0 & 5 & 8 \\ 0 & 0 & 7 \end{bmatrix}.$$

Now,

$$Ly = \begin{bmatrix} 2 & 0 & 0 \\ 3 & 5 & 0 \\ 4 & 8 & 7 \end{bmatrix} \begin{bmatrix} y_1 \\ y_2 \\ y_3 \end{bmatrix} = \begin{bmatrix} 0 \\ -160 \\ -452 \end{bmatrix}.$$

Solving for y by algebra, we get

$$y_1 = 0, \quad y_2 = -32, \quad y = -28.$$

Now we solve for x,

$$L^T x = \begin{bmatrix} 2 & 3 & 4 \\ 0 & 5 & 8 \\ 0 & 0 & 7 \end{bmatrix} \begin{bmatrix} x_1 \\ x_2 \\ x_3 \end{bmatrix} = \begin{bmatrix} 0 \\ -32 \\ -28 \end{bmatrix};$$

we get,

$$x_3 = -4, \quad x_2 = 0, \quad x_3 = 8.$$

2.4   EIGENVALUES AND EIGENVECTORS

When solving free vibration or dynamic problems, we encounter a system of equations that can be arranged as

$$[A]\{x\} = \{0\}.$$

This type of system of linear equations is referred as "homogeneous," meaning that a unique solution exists as long as the determinant of the matrix $[A]$ is not zero. This type of problem is called an eigenvalue problem, commonly seen in dynamic problems such as buckling, vibration, and electrical systems. The solution to eigenvalue problems is non-unique, meaning that many solutions exist and we are interested in a solution that is non-trivial. Rearrange the above equation as

$$([A] - \lambda[I])\{x\} = \{0\},$$

where the system matrix is $[A]$ and the identity matrix is $[I]$ with the same dimension as $[A]$. The unknown vector $\{x\}$ is called the eigenvector and "λ"s are called eigenvalues. Each eigenvalue has its own eigenvector. In free vibration problems, eigenvalue represents the frequency of vibration and the corresponding eigenvector refers to the mode shape or vibration mode. Eigenvalue problems can be found in many applications in diverse fields such as in engineering (especially dynamics and control), biology, environmental science, economics, psychology, and many other areas. The following example describes the procedures involved in solving eigenvalues and eigenvectors.

Example 2.4   For the matrix $[A]$ given below, find the eigenvalues and eigenvectors:

$$A = \begin{bmatrix} -5 & 2 \\ 2 & -2 \end{bmatrix}.$$

Solution: First, eigenvalues need to be determined using the equation, $[A]\{x\} = \lambda\{x\}$. Rewriting using the given matrix $[A]$ as

$$Ax = \begin{bmatrix} -5 & 2 \\ 2 & -2 \end{bmatrix} \begin{bmatrix} x_1 \\ x_2 \end{bmatrix} = \lambda \begin{bmatrix} x_1 \\ x_2 \end{bmatrix}.$$

In terms of components,

$$-5x_1 + 2x_2 = \lambda x_1$$
$$2x_1 - 2x_2 = \lambda x_2.$$

Transferring the terms on the right to the left, we get

$$(-5 - \lambda) x_1 + 2x_2 = 0$$
$$2x_1 + (-2 - \lambda) x_2 = 0.$$

This can be written in matrix notation

$$(A - \lambda I)x = 0.$$

This is a *homogeneous* linear system. By Cramer's theorem, the nontrivial solution (for an eigenvector of A) if and only if its coefficient determinant is zero, that is,

$$D(\lambda) = \det(A - \lambda I) = \begin{vmatrix} -5 - \lambda & 2 \\ 2 & -2 - \lambda \end{vmatrix} = (-5 - \lambda)(-2 - \lambda) - 4 = \lambda^2 + 7\lambda + 6 = 0$$

we call $D(\lambda)$ the **characteristic determinant** or, if expanded, the **characteristic polynomial**, and $D(\lambda) = 0$ the **characteristic equation** of A.

The solutions to the above quadratic equation are $\lambda_1 = -1$ and $\lambda_2 = -6$. These are the eigenvalues of A.

Eigenvector of A corresponding to λ_1. This vector is obtained from (2) with $\lambda = \lambda_1 = 1$, that is,

$$-4x_1 + 2x_2 = 0$$
$$2x_1 - x_2 = 0.$$

A solution is $x_2 = 2x_1$ as we see from either of the two equations, but we need only one of them. This determines an eigenvector corresponding to up to a scalar multiple. If we choose $x_1 = 1$, we obtain the eigenvector

$$x_1 = \begin{bmatrix} 1 \\ 2 \end{bmatrix}.$$

Check:

$$Ax_1 = \begin{bmatrix} -5 & 2 \\ 2 & -2 \end{bmatrix} \begin{bmatrix} 1 \\ 2 \end{bmatrix} = \begin{bmatrix} -1 \\ -2 \end{bmatrix} = (-1) x_1 = \lambda_1 x_1.$$

Eigenvector of A corresponding to λ_2. For $\lambda = \lambda_1 = 2$, Equation (2) becomes

$$x_1 + 2x_2 = 0$$
$$2x_1 + 4x_2 = 0.$$

A solution is $x_1 = -x_1/2$ with arbitrary x_1. If we choose $x_1 = 2$, we get $x_2 = -1$. Thus, an eigenvector of A corresponding to $\lambda_2 = 6$ is

$$x_2 = \begin{bmatrix} 2 \\ -1 \end{bmatrix}.$$

Check:

$$Ax_2 = \begin{bmatrix} -5 & 2 \\ 2 & -2 \end{bmatrix} \begin{bmatrix} 2 \\ -1 \end{bmatrix} = \begin{bmatrix} -12 \\ 6 \end{bmatrix} = (-6) x_2 = \lambda_2 x_2.$$

For the matrix in the intuitive opening example at the start of this section, the characteristic equation is

$$\lambda^2 - 13\lambda + 30 = (\lambda - 10)(\lambda - 3) = 0.$$

The eigenvalues are given as $\{10, 3\}$.

Corresponding eigenvectors are $[3\ 4]^T$ and $[-1\ 1]^T$, respectively. Please try this exercise and verify the answer.

Note: The eigenvalues of a real symmetric matrix are real. In general, eigenvalues represent natural frequencies of a dynamic system and may be related to resonance. Once the natural frequencies of a system coincide with natural frequencies, then the system will lead to resonance condition, which should be avoided.

2.5   NUMERICAL INTEGRATION

In finite element analysis, numerical integration is extensively used to calculate element matrices. Numerical integration formulas such as Simpson's rule or trapezoidal rule, often assume equally spaced data and have limited applicability and accuracy when used in finite element analysis. In a majority of finite element equations, the *Gauss Quadrature* has become the accepted numerical integration scheme. In general, an integral is evaluated as

$$I = \int_a^b f(x)dx = \sum_{k=1}^n W_k f(x_k),$$

where the x_k are the sampling points (Gauss points), and the w_k are the weights. Among various Gauss quadrature formulas, the Gauss-Legendre quadrature is most popular for finite element analysis. The weights and sampling points are based on Legendre polynomial. In Gauss-Legendre numerical integration, the limits of integration have to be between -1 to $+1$. The integrals in most finite element applications are of the form

$$I = \int_{-1}^1 f(\xi)d\xi = \sum_{k=1}^n W_k f(\xi_k),$$

where ξ is a non-dimensional coordinate which relates to the coordinate "x" in order to change the limits of integration from $(a$ to $b)$–$(-1$ to $+1)$. The relation is,

$$x = \frac{(a+b)}{2} + \frac{(b-a)}{2}\xi$$

and

$$dx = \frac{(b-a)}{2}d\xi.$$

The above equations can be easily extended to two-dimensional (2D) and three-dimensional (3D) problems. Table 2.1 shows the sampling points and weighting factors for Gauss-Legendre formula.

Table 2.1: Sampling points and weighting factors for Gauss-Legendre formula

Points	2	3
Weighting Factors (wi)	$W_1 = 1.00000000$ $W_2 = 1.00000000$	$W_1 = 0.55555556$ $W_2 = 0.88888889$ $W_3 = 0.55555556$
Sampling Points (λ_i)	$\lambda_1 = -0.577350269$ $\lambda_2 = 0.577350269$	$\lambda_1 = -0.77459669$ $\lambda_2 = 0.0$ $\lambda_3 = 0.77459669$

Example 2.5   Evaluate the following integral analytically as well as numerically and compare the results.

$$I = \int_3^7 \left(2x^3 - 5x + 10\right)\, dx.$$

Solution: Analytical

$$I = \int_3^7 \left(2x^3 - 5x + 10\right)\, dx$$

$$I = \frac{x^4}{2} - \frac{5x^2}{2} + 10x \Big|_3^7$$

$$I = 1100.$$

Solution: Numerical Integration using two-point Gauss-Legendre formula

$$x = \frac{b+a}{2} + \frac{b-a}{2}\lambda$$

$$x = \frac{7+3}{2} + \frac{7-3}{2}\lambda$$

$$x = 5 + 2\lambda; \qquad dx = 2\, d\lambda.$$

Therefore,

$$I = \int_{-1}^{1} 2\left\{2(5+2\lambda)^3 - 5(5+2\lambda) + 10\right\}\, d\lambda.$$

Then,

$$f(\lambda) = 2\left\{2(5 + 2\lambda)^3 - 5(5 + 2\lambda) + 10\right\}$$
$$I = w_1 f(\lambda_1) + w_2 f(\lambda_2)$$

$$I = 1.0 \times f(-0.577350269) + 1.0 \times f(0.577350269)$$
$$I = 1.0 \times 2\left[113.7157178 - 19.2264973 + 10\right]$$
$$+ 1.0 \times 2\left[466.2842821 - 30.77350269 + 10\right]$$
$$I = 1100.$$

2.6   MATRIX ALGEBRA USING MATLAB TOOLBOX

MATLAB is a mathematical software package that has gained a very favorable reputation and is widely used in engineering education. MATLAB is a software tool that can be used to perform many types of interactive numerical computations, data analysis and graphics. By using its built-in functions (subroutines) related to vectors and matrices, we can easily and quickly develop computer programs for computational analysis. Table 2.2 provides an overview of the MATLAB commands related to matrix algebra.

Table 2.2: MATLAB Commands for Matrix Algebra

ARITHMETIC OPERATORS	MATRIX OPERATIONS
"+" Addition	• **Dot Operator**
"-" Subtraction	o Element to element operation
"*" Multiplication	o ".*" Elemental Multiplication
"/" Division	o "./" Elemental Division
"^" Power	o ".^" Elemental Power
MATRICES AND ARRAYS	**LINEAR EQUATION SOLUTIONS**
o The element in row i and column j of A is denoted by "A(i,j)"	• The backslash,"\", can be used with matrices to solve for Ax=B.
o Matrix creation	• A\B = inv(A)*B
• Surround values with brackets "[ ]"	• x = inv(A)*B
• Separate values with spaces or commas	• Solution "x" using Gaussian Elimination for Ax=B.
• Use the semicolon to designate rows	• X = A\B
• A = [1,2,3,4;5,6,7,8]	• Solution "x" using Cholesky's Decomposition for Ax=B.
• Determinant of a matrix [A] can be found by simply using the command – det(A)	• [l, u] = lu(A)
	• X = inv(u)*inv(l)*B
	EIGENVALUES/EIGENVECTORS SOLUTIONS
	• Eigenvalues (eVal) and eigenvectors (eVec) of a matrix [A] can be found using,
	• [eVal, eVec] = eig(A)

2.7   EXERCISE PROBLEMS

2.1.   Given the matrices $[A]$ and $[B]$ below, perform the following operations, (i) $[A][B]$; (ii) $[B][A]$; (iii) $[A]^T$; (iv) $[B]^T$; (v) $[A]^{-1}$; and (vi) $[B]^{-1}$:

$$[A] = \begin{bmatrix} 1 & 2 \\ 3 & 0 \end{bmatrix} \quad \text{and} \quad [B] = \begin{bmatrix} 5 & -1 \\ 0 & 2 \end{bmatrix}.$$

2.2.   Given the matrices $[A], [B], [C]$, and $[D]$ below, perform the following operations, (i) $[A] * [B]$ (ii) $[B] * [A]$; and (iii) $([A] - [C]) * ([B] + [D])$:

$$[A] = \begin{bmatrix} 7 & 8 \\ 43 & 12 \\ 6 & 2 \end{bmatrix} [B] = \begin{bmatrix} -2 & 13 & 43 \\ 3 & 0 & -9 \end{bmatrix} [C] = \begin{bmatrix} -7 & 8 \\ 32 & 23 \\ -6 & 2 \end{bmatrix}$$

$$[D] = \begin{bmatrix} 2 & 1 & 0 \\ 8 & 3 & -12 \end{bmatrix}.$$

2.3.   Given the matrix $[A]$, find $[A]^{-1}$ by hand and compare your answer with MATLAB.

$$[A] = \begin{bmatrix} 3 & 5 & 17 \\ -2 & 3 & 1 \\ 8 & -8 & 4 \end{bmatrix}.$$

2.4.   Solve the following set of equations (a) using the Gaussian method and (b) using the LU decomposition method:

$$2x_1 - 3x_2 = 1$$
$$5x_1 + x_2 = 2.$$

2.5.   Describe engineering applications where you need to find eigenvalues/vectors and explain why. Determine all eigenvalues and corresponding eigenvectors of the following matrix:

$$[A] = \begin{bmatrix} -3 & 2 \\ 6 & -4 \end{bmatrix}.$$

2.6.   Determine all eigenvalues and corresponding eigenvectors for the matrix below and compare your answer with MATLAB:

$$[A] = \begin{bmatrix} 3 & -10 & 0 \\ -10 & 0 & 30 \\ 0 & 30 & -27 \end{bmatrix}.$$

2.7. Solve the following system of equations using (i) Gaussian method; (ii) *LU* decomposition method; and (iii) inverse matrix method by hand and compare your answers with MATLAB:

$$x + y + z = 1$$
$$3x - 18y = 20$$
$$9x + 2y - 15z = 3.$$

2.8. Evaluate the integral given below analytically using Gauss-Legendre two and three point formulae.

$$I = \int_3^7 (2x_3 - 5x + 10) \, dx.$$

2.9. Evaluate the given integral analytically and using Gauss-Legendre two- and three-point formulae:

$$I = \int_{-2}^8 (3x^4 + x^2 - 7x + 10) \, dx.$$

2.10. Give two examples/applications where eigenvalues/eigenvectors are important to study. Determine all the eigenvalues/eigenvectors for the following system of equations:

$$\left[-\omega^2 m \begin{bmatrix} 1 & 0 \\ 0 & 1 \end{bmatrix} + k \begin{bmatrix} 2 & -1 \\ -1 & 2 \end{bmatrix} \right] \begin{Bmatrix} x_1 \\ x_2 \end{Bmatrix} = \begin{Bmatrix} 0 \\ 0 \end{Bmatrix}.$$

2.11. Determine all the eigenvalues/eigenvectors for the following matrix by hand and compare your answers with MATLAB:

$$[A] = \begin{bmatrix} 3 & -10 & 0 \\ -10 & 0 & 30 \\ 0 & 30 & -27 \end{bmatrix}.$$

CHAPTER 3

Finite Element Analysis–Basics

After reading this chapter, you will be able to:

- define FEA in engineering;

- explain various methods of FE formulations;

- explain the generic FE procedure;

- explain various finite elements for modeling;

- explain symmetry, mesh refinement and coordinate systems in FEA; and

- verify FEA results.

3.1 OVERVIEW

This chapter discusses various formulations in deriving finite element equations, and a generic procedure for the finite element method. Various finite elements that can be used to model using FEM are presented. The topics of symmetry and mesh refinement as well as coordinate systems are briefly described. Finally, verification of analysis results which is an important aspect of FEM is also presented.

3.2 FEM FORMULATIONS

There are many approaches for formulating the finite element analysis problems that will result in solving a system of equations of the form $[A]\{x\} = \{b\}$. Here, some common approaches such as direct formulation, minimum total potential energy formulation and weighted residual formation are briefly described.

3.2.1 DIRECT FORMULATION

This formulation is based on the principle of engineering science and uses balance of forces, and applies statics to generate the finite element equations. Usually, it is easy to apply direct formulation to one-dimensional (1D) problems and it provides a clear physical insight into finite element analysis. The direct method of finite element analysis is demonstrated using an example, as shown below.

Consider the free body diagram of a typical element, as shown in Fig. 3.1. The element has two nodes (i and j) connected through a spring with stiffness (k_e). Assume that the coordinate system (x) is going from left (node i) to right (node j). The forces acting at the nodes are denoted as f_i and f_j. The unknown nodal displacements at nodes i and j are u_i and u_j. Now, let's develop a relationship between the nodal displacements (u_i and u_j) and the element forces (f_i and f_j).

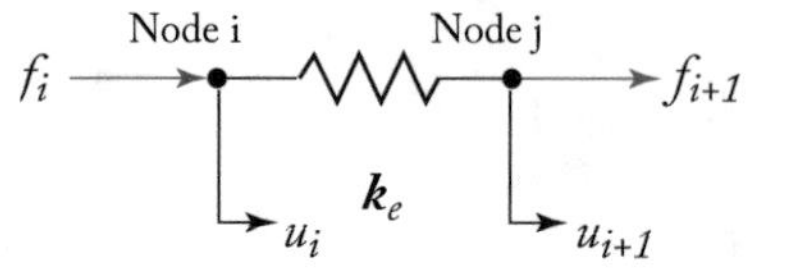

Figure 3.1: Spring element (e) connected by node i and node j.

For equilibrium, the sum of the forces acting on the element must be zero. This equals to

$$f_i + f_{i+1} = 0.$$

Now, we can write the relations, between the element forces and the displacements using (force = stiffness × displacement) as

$$f_i = k_e \, (u_i - u_{i+1})$$
$$f_{i+1} = k_e \, (u_{i+1} - u_i).$$

Rearranging the above two equations as

$$\begin{bmatrix} k_e & -k_e \\ -k_e & k_e \end{bmatrix} \begin{Bmatrix} u_i \\ u_{i+1} \end{Bmatrix} = \begin{Bmatrix} f_i \\ f_{i+1} \end{Bmatrix}$$

$$\underbrace{\text{Stiffness}} \quad \underbrace{\text{Displacement}}_{\text{vector}} \quad \underbrace{\text{Force}}_{\text{vector}}$$

or

$$[k] \, \{u\} = \{f\},$$

where $[k]$ is the element stiffness matrix, $\{u\}$ is the vector of nodal displacements or DOF (degrees of freedom) associated with the element, and the vector $\{f\}$ represents elemental forces. The above relation is called the element equilibrium equation and can be applied to any generic element. Sometimes the above equation can be used with a superscript "e" to represent elemental equations.

The direct formulation is applicable to simple problems and is helpful in understanding the underlying concepts. Complex problems can't be solved using the direct formulation and require other approaches.

3.2.2 THE MINIMUM TOTAL POTENTIAL ENERGY FORMULATION

This formulation is based on the total potential energy of the system and is mainly used for solid mechanics problems. For a given solid body, the strain energy is stored in the material as an elastic energy resulting from deformation of the body due to external applied loads. In general,

Total Potential Energy (π) = Strain Energy (U) + Work Potential (W).

The minimum total potential energy principle states that for a stable system, the displacement at the equilibrium position occurs such that the value of the system's total potential energy is a minimum.

Consider a 1D system, where the strain energy for a solid body is given by,

$$U = \frac{1}{2} \int \sigma^T \varepsilon \, dv,$$

where the stress is σ, strain is ε, and v is the volume of the body.

The work exerted by the applied forces (f_i) through the displacements (u_i) is given by,

$$W = - \int f_i u_i \, dv.$$

For a linear elastic body, the stress (σ) and the strain (ε) are related through Hooke's law as $\sigma = E\varepsilon$, then the potential energy is given as,

$$\pi = \int \frac{E\varepsilon^2}{2} \, dv - \sum f_i u_i,$$

where $\sum f_i u_i$ assumes more than one force is acting on the body.

Then, we use the minimization principle by taking the differential as

$$\frac{\partial \pi}{\partial u_i} = \frac{\partial}{\partial u_i} \sum_{i=1}^{n} U_i - \frac{\partial}{\partial u_i} \sum_{i=1}^{m} f_i u_i.$$

Noting for an axially loaded element of length (l) with two nodes i and $i + 1$, as shown in Fig. 3.2, the strain is calculated as, $\varepsilon = \frac{u_{i+1} - u_i}{l}$, and $U_e = \frac{1}{2} E \int \varepsilon^2 \, dv$, which is then, $U_e = \frac{E}{2l}(u_{i+1} - u_i)^2$ gives rise to

$$\left\{ \begin{array}{c} \frac{\partial U_e}{\partial u_i} \\ \frac{\partial U_e}{\partial u_{i+1}} \end{array} \right\} = \left[\begin{array}{cc} k & -k \\ -k & k \end{array} \right] \left\{ \begin{array}{c} u_i \\ u_{i+1} \end{array} \right\}, \quad \text{where} \quad k = \frac{E}{l}.$$

Knowing that, $\frac{\partial}{\partial u_i}(f_i u_i) = f_i$, and $\frac{\partial}{\partial u_{i+1}}(f_{i+1} u_{i+1}) = f_{i+1}$, we then get the relation as

$$\left\{ \begin{array}{c} f_i \\ f_{i+1} \end{array} \right\} = [k] \left\{ \begin{array}{c} u_i \\ u_{i+1} \end{array} \right\}.$$

This formulation is applicable for simple solid mechanics and structural problems, and not helpful for solving 2D/3D problems.

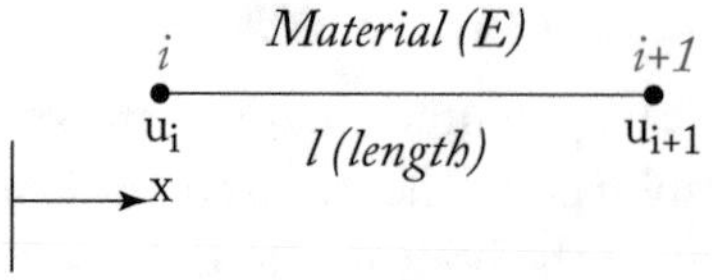

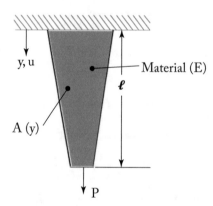

Figure 3.2: An axial bar element connected by node i and node $i + 1$.

3.2.3   WEIGHTED RESIDUAL FORMULATION

This formulation is based on assuming an approximate solution to the governing differential equations. Many simple engineering problems such as beam bending or heat conduction in a solid can be described by differential equations. The differential equation along with boundary conditions is called the boundary value problem. An example of a simple boundary value problem for a uniaxial bar subjected to a force, P at the end $(y = l)$ along its axis is shown in Fig. 3.3.

Figure 3.3: A non-uniform bar subjected to an applied load P.

Differential equation: $EA(y)\frac{du}{dy} - P = 0$.
Boundary condition: $u(y = 0) = 0$ because the displacement u at the fixed end is zero.

The solution $u(y)$ of the above differential equation needs to satisfy the differential equation at all points along the bar as well as the boundary condition. When the geometry is simple with uniform properties, it is easy to solve the differential equation analytically. But when the geometry is complicated, it is not trivial to solve for $u(y)$ analytically. So we need to solve the differential equation numerically, using an approximate solution for $u(y)$.

Usually, the assumed solution is not an exact solution, some residuals or errors will result after substituting the solution into the differential equation. The weighted residual method requires that the error be minimized over the domain (some points or intervals).

To demonstrate the weighted residual formulation, consider the example of a non-uniform bar as shown in Fig. 3.3 which is subjected to an axial force P at one end while the other end is fixed. Let us assume the solution for $u(y)$ as follows:

$$u(y) = C_1 y + C_2 y^2 + C_3 y^3,$$

where C_1, C_2, and C_3, are constants that need to be determined. Since the assumed displacements will not satisfy the differential equation exactly, there will be a residual function $R(y)$ which is given as

$$R(y) = EA(y)\frac{du}{dy} - P.$$

Our goal is to minimize the residual function $R(y)$ as much as possible. There are various methods to do this, and these are described below.

Collocation Method: In this method, the residual function $R(y)$ is forced to be zero at many points as there are unknown constants in the assumed displacement. For the assumed solution, $u(y)$ has three unknown constants (C_1, C_2 and C_3), we have to force the error or residual function at three arbitrary points. Usually these points are chosen equally along the domain.

Subdomain Method: In this method, the integral of the residual function $R(y)$ over some selected subintervals is forced to be zero, keeping in mind that the number of subintervals chosen must equal the unknown constants in the assumed displacement:

$$\int_a^b R(y)\, dy = 0.$$

For the assumed solution, $u(y)$ has three unknown constants (C_1, C_2, and C_3), and we have three integrals with different limits (a and b). For example, 1st integral may have limit from 0–$l/3$; 2nd integral limits are $l/3$–$2l/3$, and the third integral may have limits of $2l/3$–l. Solving these three integrals will give the solution to the constants C_1, C_2, and C_3.

Galerkin Method: In this method, the residual function $R(y)$ when multiplied to some weighting function $W(y)$ is orthogonal over the interval of the domain and should be equal to zero. This is given as

$$\int_a^b W_i(y)R\, dy = 0.$$

The weighting functions $W_i(y)$ are chosen to be a part of the assumed solution. For example, the assumed solution, $u(y) = C_1 y + C_2 y^2 + C_3 y^3$, has three unknown constants (C_1, C_2, and C_3), we have to select three weighting functions as $W_1 = y$; $W_2 = y^2$; and $W_3 = y^3$. With this selection, we should be able to solve for the three unknown constants C_1, C_2, and C_3.

Least Squares Method: In this method, the error is minimized with respect to the unknown constants in the assumed solution, according to the relationship

$$\text{Minimize} \int_a^b R^2 \, dy = 0.$$

Depending on how many unknown constants are in the assumed solution, the above equation will lead to as many equations to solve uniquely for the unknown constants.

3.3   GENERIC FEM PROCEDURE

In the finite element method (FEM), the actual structure/solid is discretized using finite elements. These elements are interconnected at "nodes" shared with the other elements. Each element has a specific shape (a line, a triangle or a rectangle) and a specific number of nodes. The unknowns at each node are the field variables (displacement, temperature, velocity, or pressure). These variables are also called DOF (degrees of freedom). After discretization of the structure, boundary conditions (displacements) and applied loads are specified. Global equations are obtained by assembling elemental equations, and then solved for the unknown displacements given the forces and boundary conditions. From the displacements at the nodes, the stresses and strains are calculated in each element.

In applying FEM to solve real practical problems, the design analyst/engineer is faced with questions such as the following.

- How to model the problem using finite elements?

- What kind of elements and how many elements should be used in the model?

- How should the boundary conditions and loads be specified?

- How to interpret the results?

The answers to the above questions will be based on formal training and experience. In general, FEM involves three basic phases. These are briefly described below.

- *Pre-processing Phase*

    - Create and discretize the solution domain into specific finite elements with nodes and elements.

    - Develop equations for an element with specific behavior.

    - Construct the global matrix by assembling the element matrices.

    - Apply loading, boundary conditions, and initial conditions.

- *Solution Phase*

  - Solve a set of linear or nonlinear algebraic equations simultaneously to obtain nodal results (displacement values in stress analysis or at different nodal temperatures in heat transfer).

- *Post-processing Phase*

  - Obtain other quantities such as stresses, heat fluxes derived from nodal variables, and review results for accuracy and correctness.

3.4 FINITE ELEMENT MODELING TECHNIQUES

3.4.1 TYPES OF FINITE ELEMENTS

There is no unique way of modeling using finite elements for specific engineering problems. Different types of elements can be used to model the same physical problem. An important aspect is that the user/engineer needs to understand the capabilities and limitations of various elements in modeling the physical problem. Figure 3.4 shows various types of elements. 1D elements are bar and beam elements and 2D elements are either triangular or quadrilateral in shape and are

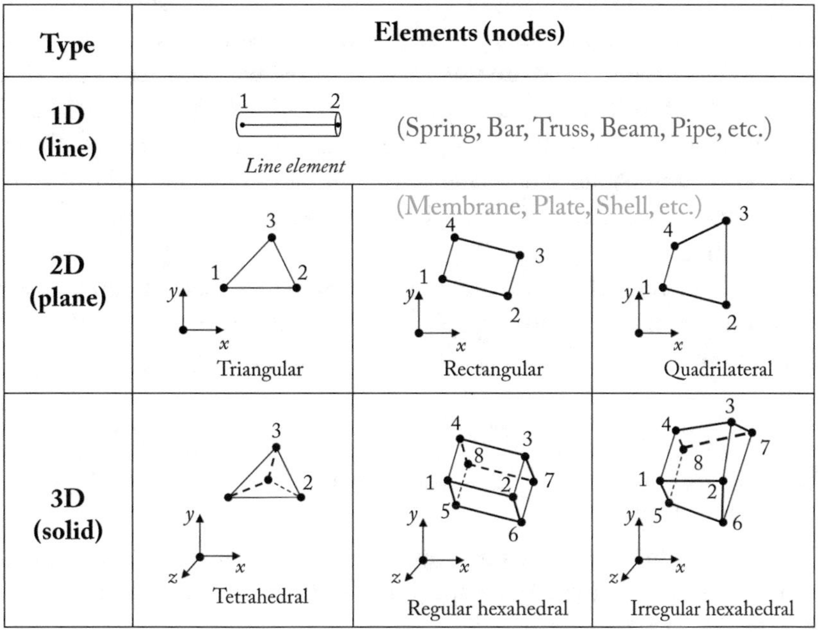

Figure 3.4: Various types of finite elements used to model in FEA.

used for plane solid/plate geometries. 3D elements are modeled using tetrahedral/hexahedral elements. In addition, plate/shell finite elements can be used to model thin sheets/plates and curved surfaces. These types of elements will have both in-plane and out-of-plane behavior to model thin structures.

Most of the elements shown in Fig. 3.4 represent linear elements (two-node bar, three-node triangular, four-node quadrilateral, eight-node hexahedral). However, to model sharply curved geometries, higher order elements can be used. These include quadratic elements (three-node bar, six-node triangular, eight-node quadrilateral, twenty-node hexahedral) and cubic elements (four-node bar, nine-node triangular, twelve-node quadrilateral, thirty two-node hexahedral). In these higher-order elements, each edge is represented by either three nodes for quadratic elements or four-nodes for cubic elements. In general, higher-order elements are more accurate than linear elements.

3.4.2   COORDINATE SYSTEMS

In FEM, different coordinate systems are used as a reference in order to specify different quantities easily and meaningfully. In the global coordinate system, the node location, element orientation, the loads and the boundary conditions are specified conveniently. Also, the solution (nodal variables) is generally represented in the global coordinate system. A local coordinate system is very convenient to attach to an element in order to simplify the derivation of element stiffness matrix through algebraic manipulations. A natural coordinate system is a local coordinate system that uses a dimensionless parameter whose limits are within -1 to $+1$. Therefore, natural coordinates are dimensionless and are usually represented with respect to the element rather than the global coordinate system. The use of a natural coordinate system simplifies the integration when evaluating a stiffness matrix using numerical integration formulas. Certain types of finite elements known as isoparametric elements use natural coordinates and play a crucial role in modeling curved boundaries. In general, global, local, and natural coordinate systems are related through transformation.

Consider a 1D system with a global coordinate system (X), local coordinate system (x) and a natural coordinate system (ξ), as shown in Fig. 3.5. The relationship among various coordinate systems is given by

$$X = X_i + x$$
$$\xi = (2x/l) - 1.$$

Using these various coordinate systems, the element shape functions and the corresponding finite element formulations can be derived. The shape functions using these different coordinate systems are described in the next chapter.

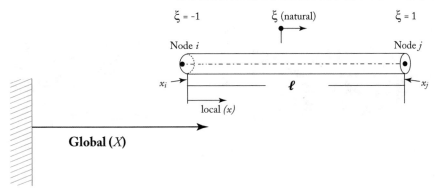

Figure 3.5: Different coordinate systems—global (X), local (x), and natural (x).

3.4.3   USING SYMMETRY

In general, if a given physical problem is symmetric with respect to geometry, material properties, applied loads, and boundary conditions, then exploiting the symmetry in FEA will greatly reduce the computational effort. Different types of symmetry exist (axisymmetry; rotational symmetry; or planar or reflective symmetry; repetitive or translational symmetry) in different problems. If the physical problem is not symmetric and loads/boundary conditions are being applied symmetrically, then the whole geometry needs to be analyzed with the FEM. Figure 3.6a shows an example of an elastic plate with holes subjected to tensile loadings as well as a symmetric model (one-half model of the problem). Similarly, the symmetry model (one-quarter model of the problem) for an example of fluid around pipe is shown in Fig. 3.6b. In general, the vertical line of symmetry will have zero displacements in the horizontal direction, and the horizontal line of symmetry will have zero displacements in the vertical direction.

3.4.4   MESH REFINING (FREE MESHING VS. MAPPED MESHING)

Mesh refinement is needed when solving problems with geometric discontinuities such as curves, holes, cracks, fillets, and so on, where there is a stress concentration. In order to predict the stress concentrations, the mesh needs to be refined around geometric discontinuities as shown in Fig. 3.6a. There are two ways of creating meshing, one is *free meshing* in which the user provides a general guideline, and the preprocessing program will mesh accordingly. Usually, free meshing is used for modeling complex geometries and may involve more than one type of element (triangular or quadrilateral, and so on). In general, free mesh has no element restrictions, does not follow any pattern, and is suitable for complex shaped objects.

On the other hand, in *mapped meshing* (mesh refining), the user provides detailed instructions on how the mesh should be created. In general, more user action is required and the user has to control the size and shape of elements covering the computational domain. Mapped meshing is also restricted to quadrilateral element shapes, typically follows a regular pattern, and is suitable

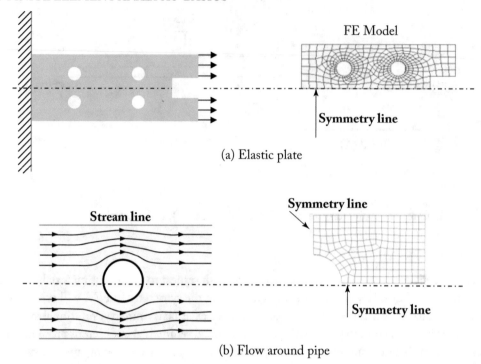

(a) Elastic plate

(b) Flow around pipe

Figure 3.6: (a) Elastic plate subjected to tensile loading and (b) flow around a pipe example for fluid analysis.

for regularly shaped objects. Figure 3.7 shows an example of mapped mesh as well as free mesh created for a bracket.

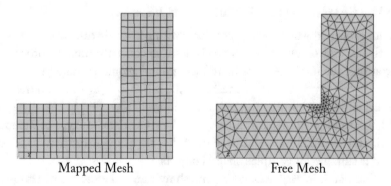

Mapped Mesh   Free Mesh

Figure 3.7: Mapped mesh and free mesh in ANSYS software.

3.5   VERIFICATION OF FEA RESULTS

In using finite element analysis as a design tool, engineers must understand the limitations of the finite element analysis method, which is a numerical method. Especially, when using general purpose software like ANSYS, there are various sources of error that can contribute to incorrect results. These can be corrected as follows.

Checking Inputs: Geometry dimensions, materials properties, appropriate boundary conditions, and applied loads.

Selecting correct types of elements: Appropriate elements and their underlying theory for the governing behavior to the problem at hand.

Element Properties: When modeling the geometry, selection of inappropriate element shape and size will affect the results. When meshing the geometry, the element shape and size should be keep in mind for the analysis.

Applying correct boundary conditions: Once the finite element model is built, applying correct displacement boundary conditions is very important in order to have a unique solution. The finite element model needs to be constrained, both translation/rotation in order to restrain the rigid body motions, otherwise the finite element solution is not possible due to singularity of the matrix system of equations.

Checking Results: Usually, the equilibrium (force/momentum) and the energy balance equations should be applied to different parts of the geometry to see if any physical laws are being violated. For example, in static problems, the equilibrium of forces acting on the free body diagram should be satisfied. Mostly, the checking of results involves having some knowledge about the problem, using intuition and sometimes, gut feeling will help to identify inaccurate results from analysis.

3.6   EXERCISE PROBLEMS

3.1.   Explain the basics of finite element analysis using direct formulation.

3.2.   Discuss various finite element formulations.

3.3.   What are various methods available to satisfy the weighted residual formulation?

3.4.   Explain how symmetry can be used to simplify the analysis model.

3.5.   Explain when you need to carry out mesh refinement.

3.6.   Find a relation between the coordinate systems shown in Fig. 3.8.

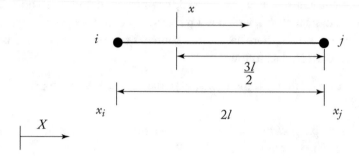

Figure 3.8: Exercise Problem 3.6.

3.7.   Find a relation between the coordinate systems shown in Fig. 3.9.

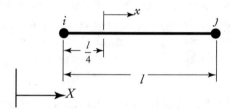

Figure 3.9: Exercise Problem 3.7.

3.8.   Find the shape functions for a linear element, shown in Fig. 3.10, using the local coordinate ξ.

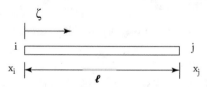

Figure 3.10: Exercise Problem 3.8.

3.9.   How do you verify finite element analysis results?

CHAPTER 4 title

C H A P T E R   4

1D Finite Element Analysis

After reading this chapter, you will be able to:

- understand 1D problems in engineering;

- define and find shape functions for 1D problems;

- know various types of elements for 1D problems;

- understand and analyze stress analysis problems;

- understand and analyze heat transfer problems;

- understand and analyze fluid problems; and

- understand free vibration analysis.

4.1   OVERVIEW

The finite element process is a systematic methodology for solving design problems for stress analysis, heat transfer analysis, fluid dynamic analysis, free vibration analysis and others. 1D elements represent engineering problems involving only one dimension. The derivation of 1D shape functions is also presented. This chapter introduces procedures involved in applying finite element analysis for 1D problems involving stress, heat transfer, and fluid with examples.

4.2   1D ELEMENTS

One-dimensional (1D) elements such as axial bar, horizontal beam, and heat transfer through a finned tube represent engineering problems with only one dimension. The fundamental concept in FEM is that a continuous function is approximated through a discrete model. The continuous function is divided into elements, with end points. 1D finite elements are derived using a simple polynomial function to describe the behavior between its end points (nodes) and are usually valid for only one dimension.

4.2.1   FINITE ELEMENT MATRIX EQUATIONS

In general, FE matrix equations resulting from a formulation can be obtained using the following steps.

Step 1. Select an element to model the specific geometry/problem.

Step 2. Assume the solution for unknown variable (either displacement or temperature or any other) within the element using shape functions.

Step 3. Derive the element matrices by substituting the assumed solution into the governing differential equation. After simplifying the equations, the element matrices are obtained.

Step 4. Assemble the element matrices to get global matrices in the form $[K]\{X\} = \{F\}$.

Step 5. Solve for the unknown nodal values $\{X\}$ after applying the boundary conditions.

The next section describes the procedure for finding shape functions (step 2) for 1D elements.

4.3   SHAPE FUNCTIONS

The shape function is usually the coefficient that appears in the interpolation polynomial, and refers to element nodes of a finite element. For example, consider a 1D axial bar element with two end nodes shown in Fig. 4.1.

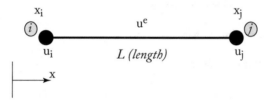

Figure 4.1: Axial bar element with two end nodes.

The displacement within the axial bar element can be represented as

$$u_{element} = u^{(e)} = s_i u_i + s_j u_j,$$

where s_i and s_j are the shape functions, and u_i and u_j are the displacements at nodes i and j, respectively.

Let's assume that the displacement u varies over the element as

$$u(x) = C_1 + C_2 x.$$

To find the constants, we need to use the element boundary conditions. These are

$$u = u_i @x = x_i$$

$$u = u_j @x = x_j.$$

Substituting the nodal values into elemental equation, results in two equations, which can be solved for the two constants:

$$u(x = x_i) = u_i = C_1 + C_2(x_i)$$
$$u(x = x_j) = u_j = C_1 + C_2(x_j).$$

Solving for two constants from the above two equations results in

$$C_1 = \frac{u_i x_j - u_j x_i}{x_j - x_i}$$

$$C_2 = \frac{u_j - u_i}{x_j - x_i}.$$

Substituting the constants in the element displacement equation for u, and comparing to the equation with shape functions, we get

$$u_{element} = u^{(e)} = s_i u_i + s_j u_j$$

$$s_i = \frac{x_j - x}{l}; \qquad s_j = \frac{x - x_i}{l}.$$

The above shape function equations can be extended to other nodal variables such as temperature, or velocity, or rotations, and others.

4.3.1   SHAPE FUNCTION PROPERTIES

- The sum of the shape functions within the element is equal to one. That is, $s_i + s_j = 1$.

- The derivative of sum of the shape functions within the element is equal to zero. That is, $\frac{d}{dx} s_i + \frac{d}{dx} s_j = 0$.

- The shape function at the corresponding node is one and everywhere else is zero, i.e., where s_i is equal to 1 at node i but zero at node j. Similarly, the shape function s_j is equal to 1 at node j but zero at node i.

The above properties of shape functions can be easily extended to multiple dimensions (either 2D or 3D). These are discussed in later chapters.

4.4   STRESS ANALYSIS

In the mechanical design process, stress analysis is considered to be an important part as many considerations can influence the design. The most important consideration in any design is that the product/component/structure should not fail during the life cycle. The failure may be due to excessive stress/strain, brittle or ductile fracture, yielding or fatigue, and depends on the material used. Through stress analysis, the design can be evaluated for robustness from failure by considering the following criteria: (i) the deflection should not exceed the maximum allowable for proper function of the product/system; (ii) the stress at every point should not exceed the maximum allowable stress for the material; and (iii) the product/system should not fail by fatigue. In addition to the above criteria, the product/system may fail in an unstable fashion or in buckling, which requires the use of a more sophisticated analysis than stress analysis.

In this chapter, a discussion on 1D stress analysis of an axial bar, heat transfer analysis, torsion analysis of a shaft, and fluid analysis is presented. The procedures for solving the same problems with ANSYS software are discussed in later chapters.

4.4.1 AXIAL BAR ELEMENT

Consider a simple example of a cylindrical rod of length l and cross-sectional area A, with a tensile force F applied along the axis in the positive x direction, as shown in Fig. 4.2.

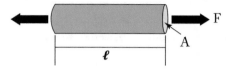

Figure 4.2: Axial bar under tensile load.

From mechanics of materials, the normal stress (σ), and the axial strain (ε), are related to the Young's modulus (E) of the material through the following relations:

$$\sigma = \frac{F}{A}; \qquad \varepsilon = \frac{\Delta l}{l}; \qquad E = \frac{\sigma}{\varepsilon}.$$

The deformation under the load is Δl. Rearranging the above equations, we can write

$$F = \frac{AE}{l}\Delta l.$$

The above equation is similar to the equation for a linear spring, which is given as

$$F = k_e\Delta x,$$

where k is the spring constant and Δx is the change in length of the spring due to the applied force F. Comparing the above two equations, we get

$$F = k_e\Delta x \quad \text{with} \quad k_e \equiv \frac{AE}{l}.$$

Now, consider a bar element of length "L" with a constant cross-sectional area "A", made from a material with modulus of elasticity "E". The bar is subjected to tensile forces "f" along its axis, as shown in Fig. 4.3.

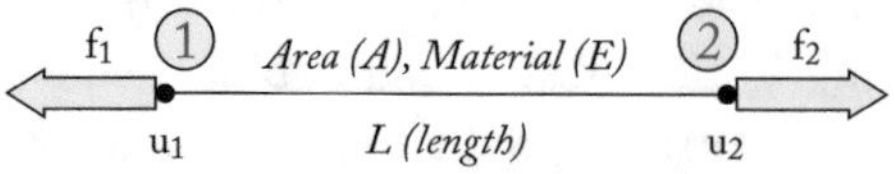

$$f_1 \;①\; \textit{Area (A), Material (E)} \;②\; f_2$$
$$u_1 \qquad\quad \textit{L (length)} \qquad\quad u_2$$

Figure 4.3: An axial bar element.

The bar element has two nodes, and is subjected to nodal forces f_1 and f_2 at nodes 1 and 2, respectively. The unknown nodal displacements are u_1 and u_2, respectively. The finite element formulation using the direct method as discussed in the previous chapter, results in the following equations:

$$\frac{EA}{L}\begin{bmatrix} 1 & -1 \\ -1 & 1 \end{bmatrix}\begin{Bmatrix} u_1 \\ u_2 \end{Bmatrix} = \begin{Bmatrix} f_1 \\ f_2 \end{Bmatrix}$$

or

$$[k]\{u\} = \{f\},$$

where $[k]$ is the axial bar element stiffness matrix, $\{u\}$ is the vector of nodal displacements or degrees of freedom (DOF) associated with the bar element, and the vector $\{f\}$ is elemental forces. Let us consider the following examples to illustrate the application of bar elements for stress analysis.

Example 4.1   A bar made from steel (Young's Modulus, $E = 30 \times 10^6$ psi), is fixed at one end and the other end is subjected an axial load as shown in Fig. 4.2. The geometric properties are, Length, $l = 10$ in, Diameter, $d = 1$ in, Cross-sectional area, $A = 0.7854$ in^2; and the applied force $F = 5,000$ lb. Model the bar using one axial bar element to find the deflection and stress in the bar.

Solution: The element stiffness, k, is calculated as

$$k = \frac{AE}{l} = \frac{\left(0.7854\,\text{in}^2\right)\left(30 \times 10^6\,\text{lb/in}^2\right)}{1}0\,\text{in} = 2.3652 \times 10^6\,\text{lb/in}^2.$$

The values for k and f can now be substituted into the *FE* matrix equation.

$$\begin{bmatrix} 2.3652 \times 10^6 & -2.3652 \times 10^6 \\ -2.3652 \times 10^6 & 2.3652 \times 10^6 \end{bmatrix}\begin{Bmatrix} u_1 \\ u_2 \end{Bmatrix} = \begin{Bmatrix} -5000 \\ 5000 \end{Bmatrix}.$$

Applying the boundary condition $u_1 = 0$ yields

$$\begin{bmatrix} 1 & 0 \\ -2.3652 \times 10^6 & 2.3652 \times 10^6 \end{bmatrix}\begin{Bmatrix} u_1 \\ u_2 \end{Bmatrix} = \begin{Bmatrix} 0 \\ 5000 \end{Bmatrix}.$$

Solving the matrix equation, the displacement u_2 is calculated to be 2.12×10^{-3} in.

Using the calculated value of u_2 and rearranging, the stress in the element is calculated as:

$$\sigma_x = E\varepsilon = \frac{E\Delta l}{l} = \frac{E(u_2 - u_1)}{l} = \frac{\left(30 \times 10^6\,\text{psi}\right)\left(2.1221 \times 10^{-6}\,\text{in} - 0\,\text{in}\right)}{10\,\text{in}}$$

$$\sigma_x = 6366\,\text{psi}.$$

Example 4.2   A compound axial member is subjected to the loads shown in Fig. 4.4. Given, $E_1 = 50$ MN/m^2, $E_2 = 100$ MN/m^2, $L_1 = 0.5$ m, $L_2 = 1$ m, $A_1 = 20$ cm^2, and $A_2 = 10$ cm^2,

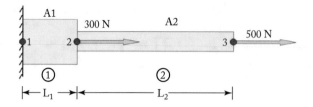

Figure 4.4: A compound axial member subjected to loads.

find (i) reaction force at points 1 (F_1) and (ii) displacements at nodes 2 and 3 (u_2 and u_3) using two bar elements model.

Solution:

Pre-processing: Number of elements $= 2$; and number of nodes $= 3$.

The stiffness matrix of each element is computed from

$$[K]^{(e)} = \frac{AE}{l} \begin{bmatrix} 1 & -1 \\ -1 & 1 \end{bmatrix}.$$

For element (1), the stiffness matrix is

$$[K]^{(1)} = \frac{0.002 \times (50 \times 10^6)}{0.5} \begin{bmatrix} 1 & -1 \\ -1 & 1 \end{bmatrix}$$

$$[K]^{(1)} = 2 \times 10^5 \begin{bmatrix} 1 & -1 \\ -1 & 1 \end{bmatrix}.$$

For element (2), the stiffness matrix is

$$[K]^{(2)} = \frac{0.001 \times (100 \times 10^6)}{1} \begin{bmatrix} 1 & -1 \\ -1 & 1 \end{bmatrix}$$

$$[K]^{(2)} = 1 \times 10^5 \begin{bmatrix} 1 & -1 \\ -1 & 1 \end{bmatrix}.$$

The global stiffness matrices for each element are

$$[K]^{1G} = 10^5 \begin{bmatrix} 2 & -2 & 0 \\ -2 & 2 & 0 \\ 0 & 0 & 0 \end{bmatrix} \qquad [K]^{2G} = 10^5 \begin{bmatrix} 0 & 0 & 0 \\ 0 & 1 & -1 \\ 0 & -1 & 1 \end{bmatrix}.$$

The final global matrices of stiffness, displacement, and load matrix, are obtained by combining the element matrices as

$$[K]^G = 10^5 \begin{bmatrix} 2 & -2 & 0 \\ -2 & 2+1 & -1 \\ 0 & -1 & 1 \end{bmatrix}; \ \{u\} = \begin{Bmatrix} u_1 \\ u_2 \\ u_3 \end{Bmatrix}; \ \{F\} = \begin{Bmatrix} 0 \\ 300 \\ 500 \end{Bmatrix}.$$

Solution: Applying the fixed boundary conditions at node 1 and applying the external forces at nodes 2 and 3, we have

$$10^5 \begin{bmatrix} 2 & -2 & 0 \\ -2 & 2+1 & -1 \\ 0 & -1 & 1 \end{bmatrix} \begin{Bmatrix} 0 \\ u_2 \\ u_3 \end{Bmatrix} = \begin{Bmatrix} F_1 \\ 300 \\ 500 \end{Bmatrix}.$$

Using the matrix partitioning to solve for u_2 and u_3, we have

$$10^5 \begin{bmatrix} 3 & -1 \\ -1 & 1 \end{bmatrix} \begin{Bmatrix} u_2 \\ u_3 \end{Bmatrix} = \begin{Bmatrix} 300 \\ 500 \end{Bmatrix}$$

$$\begin{Bmatrix} u_2 \\ u_3 \end{Bmatrix} = \begin{Bmatrix} 0.004 \\ 0.009 \end{Bmatrix} m.$$

Post-processing: Using the first row of the global matrix to find F_1, we have

$$F_1 = -2 \times 10^5 \times 0.004 = -800 \, \text{N}.$$

Check for Equilibrium: Action Forces $= -$ Reaction Forces

$$F_2 + F_3 = 300 + 500 = 800 \, \text{N} \qquad F_1 = -800 \, \text{N}.$$

4.4.2   TORSIONAL BAR ELEMENT

Consider a shaft of length "l" with a polar moment of inertia "J" made from a material with shear modulus "G". The shaft is subjected to torques "T" along its axis, as shown in Fig. 4.5.

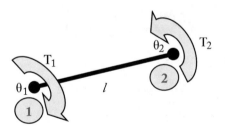

Figure 4.5: Torsional bar element.

The torsional element has two nodes, and is subjected to nodal torques T_1 and T_2 at nodes 1 and 2, respectively. The unknown nodal displacements are the twists θ_1 and θ_2. The finite element formulation results in the following equations relating the element stiffness matrix, the nodal vector of twist angles, and the elemental torques according to the equation shown below:

$$\frac{JG}{l} \begin{bmatrix} 1 & -1 \\ -1 & 1 \end{bmatrix} \begin{Bmatrix} \theta_1 \\ \theta_2 \end{Bmatrix} = \begin{Bmatrix} T_1 \\ T_2 \end{Bmatrix}.$$

The above relation for torsional bar element is similar to axial bar element, except the property GJ is similar to EI and the linear twists are similar to linear displacements.

Example 4.3　For the circular shaft subjected to torques as shown in Fig. 4.6, find (i) reaction torques at points 1 and 4 (T_1 and T_4) and (ii) angles of twist at points 2 and 3 (θ_2 and θ_3) using two finite elements. Assume, $E = 200$ GN/m², $G = 80$ GN/m², $L_1 = L_3 = 200$ mm, $L_2 = 300$ mm, $d_1 = d_3 = 10$ mm, $d_2 = 20$ mm, and $T = 5$ N.m.

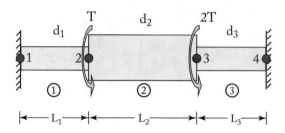

Figure 4.6: Example problem of a torsional shaft.

Solution:

Pre-processing: Number of elements = 3; and number of nodes = 4.

For circular cross section rod, $J = \pi d^4/32$; and the stiffness matrix of each element is computed from

$$[K]^{(e)} = \frac{JG}{l} \begin{bmatrix} 1 & -1 \\ -1 & 1 \end{bmatrix}.$$

So, for element (1), the stiffness matrix is

$$[K]^{(1)} = \frac{80 \times 10^9 (\pi \times 0.01^4)}{32 \times 0.2} \begin{bmatrix} 1 & -1 \\ -1 & 1 \end{bmatrix}$$

$$[K]^{(1)} = 393 \begin{bmatrix} 1 & -1 \\ -1 & 1 \end{bmatrix}.$$

For element (2), the stiffness matrix is

$$[K]^{(2)} = \frac{80 \times 10^9 (\pi \times 0.02^4)}{32 \times 0.3} \begin{bmatrix} 1 & -1 \\ -1 & 1 \end{bmatrix}$$

$$[K]^{(2)} = 4189 \begin{bmatrix} 1 & -1 \\ -1 & 1 \end{bmatrix}.$$

For element (3), the stiffness matrix is

$$[K]^{(3)} = \frac{80 \times 10^9 (\pi \times 0.01^4)}{32 \times 0.2} \begin{bmatrix} 1 & -1 \\ -1 & 1 \end{bmatrix}$$

$$[K]^{(3)} = 393 \begin{bmatrix} 1 & -1 \\ -1 & 1 \end{bmatrix}.$$

The global stiffness matrices for each element are

$$[K]^{(1G)} = \begin{bmatrix} 393 & -393 & 0 & 0 \\ -393 & 393 & 0 & 0 \\ 0 & 0 & 0 & 0 \\ 0 & 0 & 0 & 0 \end{bmatrix} \begin{matrix} \theta_1 \\ \theta_2 \\ \theta_3 \\ \theta_4 \end{matrix} \qquad [K]^{(2G)} = \begin{bmatrix} 0 & 0 & 0 & 0 \\ 0 & 4189 & -4189 & 0 \\ 0 & -4189 & 4189 & 0 \\ 0 & 0 & 0 & 0 \end{bmatrix} \begin{matrix} \theta_1 \\ \theta_2 \\ \theta_3 \\ \theta_4 \end{matrix}$$

$$[K]^{(3G)} = \begin{bmatrix} 0 & 0 & 0 & 0 \\ 0 & 0 & 0 & 0 \\ 0 & 0 & 393 & -393 \\ 0 & 0 & -393 & 393 \end{bmatrix} \begin{matrix} \theta_1 \\ \theta_2 \\ \theta_3 \\ \theta_4 \end{matrix}.$$

The final global stiffness matrix is obtained by combining the global matrices of each element

$$[K]^{(G)} = \begin{bmatrix} 393 & -393 & 0 & 0 \\ -393 & 393 + 4189 & -4189 & 0 \\ 0 & -4189 & 393 + 4189 & -393 \\ 0 & 0 & -393 & 393 \end{bmatrix} \begin{matrix} \theta_1 \\ \theta_2 \\ \theta_3 \\ \theta_4 \end{matrix}.$$

Solution: Applying the fixed boundary conditions at nodes 1 and 4 and applying the external torques at nodes 2 and 3, we have

$$\begin{bmatrix} 393 & -393 & 0 & 0 \\ -393 & 4582 & -4189 & 0 \\ 0 & -4189 & 4582 & -393 \\ 0 & 0 & -393 & 393 \end{bmatrix} \begin{Bmatrix} 0 \\ \theta_2 \\ \theta_3 \\ 0 \end{Bmatrix} = \begin{Bmatrix} T_1 \\ 5 \\ 10 \\ T_4 \end{Bmatrix}.$$

Using the matrix partitioning to solve for θ_2 and θ_3, we have

$$\begin{bmatrix} 4582 & -4189 \\ -4189 & 4582 \end{bmatrix} \begin{Bmatrix} \theta_2 \\ \theta_3 \end{Bmatrix} = \begin{Bmatrix} 5 \\ 10 \end{Bmatrix}$$

$$\begin{Bmatrix} \theta_2 \\ \theta_3 \end{Bmatrix} = \begin{Bmatrix} 0.0188 \\ 0.0194 \end{Bmatrix} \text{ rad.}$$

Post-processing: Using the first row of the global matrix to find T_1, we have

$$T_1 = -393 \times 0.0188 = -7.38 \text{ N.m.}$$

Using the last row of the global matrix to find T_4, we have

$$T_4 = -393 \times 0.0194 = -7.62 \text{ N.m.}$$

Check for Equilibrium: Action Torques $= -$ Reaction Torques

$$T_2 + T_3 = 10 + 5 = 15 \text{ N.m.} \qquad T_1 + T_4 = -7.38 - 7.62 = -15 \text{ N.m.}$$

Example 4.4   A circular shaft made from two different materials is fixed at points 1 and 3, and subjected to a torque $T = 300$ in-lb at point 2. Material A is aluminum with a shear modulus $G_A = 4.0 \times 10^6$ lb/in², and material B is steel with a shear modulus $G_B = 11.5 \times 10^6$ lb/in². Determine the angle of twist at point 2, and the reactions at nodes 1 and 3 (see Fig. 4.7).

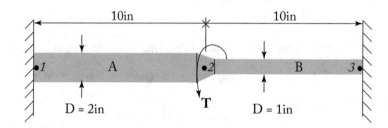

Figure 4.7: Figure of Example 4.4.

Solution:

Pre-processing: Number of elements $= 2$; and number of nodes $= 3$. For circular cross section rod, $J = \pi r^4/2$; and the stiffness matrix of each element is computed from $J = \frac{1}{2}\pi r^4$.

$$J_A = \frac{1}{2}\pi(1.0 \text{ in})^4 = 1.5708 \text{ in}^4 \qquad J_B = \frac{1}{2}\pi(0.5 \text{ in})^4 = 0.0982 \text{ in}^4$$

$$[K]^1 = \frac{J_A G_A}{l_A}\begin{bmatrix} 1 & -1 \\ -1 & 1 \end{bmatrix} \qquad [K]^2 = \frac{J_B G_B}{l_B}\begin{bmatrix} 1 & -1 \\ -1 & 1 \end{bmatrix}$$

$$\frac{J_A G_A}{l_A} = \frac{\left(1.5708 \text{ in}^4\right)\left(4.0 \times 10^6 \text{ lb/in}^2\right)}{10 \text{ in}} = 6.2832 \times 10^5 \text{ in-lb}$$

$$\frac{J_B G_B}{l_B} = \frac{\left(0.0982 \text{ in}^4\right)\left(11.5 \times 10^6 \text{ lb/in}^2\right)}{10 \text{ in}} = 1.1290 \times 10^5 \text{ in-lb}$$

$$[K]^1 = 10^5 \begin{bmatrix} 6.2832 & -6.2832 \\ -6.2832 & 6.2832 \end{bmatrix}$$

$$[K]^2 = 10^5 \begin{bmatrix} 1.129 & -1.129 \\ -1.129 & 1.129 \end{bmatrix}$$

$$[K]^G = 10^5 \begin{bmatrix} 6.2382 & -6.2382 & 0 \\ -6.2382 & 7.4122 & -1.129 \\ 0 & -1.129 & 1.129 \end{bmatrix}$$

$$10^5 \begin{bmatrix} 6.2382 & -6.2382 & 0 \\ -6.2382 & 7.4122 & -1.129 \\ 0 & -1.129 & 1.129 \end{bmatrix} \begin{Bmatrix} \theta_1 \\ \theta_2 \\ \theta_2 \end{Bmatrix} = \begin{Bmatrix} T_1 \\ T_2 \\ T_2 \end{Bmatrix}.$$

Applying B.C.'s $\theta_1 = \theta_3 = 0$ (eliminating the rows and columns corresponding to θ_1 and θ_3); and using $T_2 = 300$ in-lb:

$$10^5 \begin{bmatrix} 1/10^5 & 0 & 0 \\ -6.2382 & 7.4122 & -1.129 \\ 0 & 0 & 1/10^5 \end{bmatrix} \begin{Bmatrix} \theta_1 \\ \theta_2 \\ \theta_2 \end{Bmatrix} = \begin{Bmatrix} T_1 \\ 300 \\ T_2 \end{Bmatrix}$$

$$\boldsymbol{\theta_2 = 0.000405} \text{ rad.}$$

The reactions at the supports can be found from,

$$\{R\} = [K]^G \{\theta\} - \{T\} = \begin{Bmatrix} -254.25 \\ 0 \\ -45.75 \end{Bmatrix}.$$

Check for Equilibrium: Action Torques $= -$ Reaction Torques

$$T = T_1 + T_3$$

$$300 \text{ in-lb} = 254.25 \text{ in-lb} + 45.75 \text{ in-lb}.$$

4.5   HEAT TRANSFER ANALYSIS

Whenever there is a temperature gradient (ΔT) in a solid body, heat will flow from a high temperature side/region (T_1) to a low temperature side/region (T_2). The basic governing heat conduction equations are obtained by *Fourier law*, which states that the magnitude of the heat fluxes (heat flow/unit time) is proportional to the temperature gradient. The proportionality sign is replaced by an equal sign through a constant, k.

Assuming that the temperature varies along the x-axis (ex. thin long rod or furnace wall thickness as shown in Fig. 4.8), the heat flux is given by

$$q_x = -kA\frac{dT}{dx},$$

where k is the thermal conductivity, and is a solid material property, A is the area of cross-section of the geometry (normal to the x-axis), and q_x is the heat flux in the x-direction. The negative sign indicates that the direction of heat flux is opposite to that of the temperature gradient. The unit of heat flux is Watts, whereas the unit of thermal conductivity is W/m/°C. The same approach can be extended to multiple dimensions.

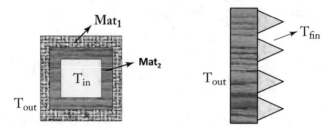

Figure 4.8: Heat transfer through a furnace wall (left) and triangular fins (right).

In addition to heat conduction, there are other modes of heat transfer which involve convection and radiation. The convection heat transfer, which can occur at the end faces of the one-dimensional body, is given by

$$Q_h = h\left(T^s - T^f\right),$$

where h is the *convection coefficient*, T^s is the temperature of the surrounding fluid, and T^f is the surface temperature of the solid. The unit of convection coefficient is W/m^2/°C.

Heat radiation, is a process following the laws of electromagnetics, in which the thermal energy is exchanged between two surfaces. The rate of radiation heat transfer is given by

$$Q_R = \sigma \varepsilon A\left(T^b - T^s\right),$$

where σ is the *Stefan–Boltzmann constant*, ε is the surface *emissivity*, A is the *surface area of the body through which heat flows*, T^b is the surface temperature of the body, and T^s is the surrounding temperature. The unit of *Stefan–Boltzmann constant* is W/m^2/°C.

4.5.1   HEAT TRANSFER ELEMENT

Consider a heat transfer element of length "L" with constant cross-sectional area "A", made from a material with thermal conductivity "k". The element is subjected to heat fluxes "q" along x-axis, as shown in Fig. 4.9.

The heat transfer element has two nodes (i and $i + 1$), and is subjected to nodal heat fluxes q_i and q_{i+1} at nodes i and $i + 1$, respectively. The unknown nodal temperatures are forces T_i and T_{i+1}. The finite element formulation results in the following equations:

$$\frac{kA}{L}\begin{bmatrix} 1 & -1 \\ -1 & 1 \end{bmatrix}\begin{Bmatrix} T_i \\ T_{i+1} \end{Bmatrix} = \begin{Bmatrix} q_i \\ q_{i+1} \end{Bmatrix}.$$

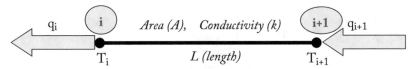

Figure 4.9: Heat transfer finite element.

The U-factor is defined as $U = k/L$, which represents thermal transmission per unit area that has units of Btu/hr.ft^2.°F. Sometimes, U is also called as the reciprocal of thermal resistance. Therefore, the above equations can be written as

$$UA \begin{bmatrix} 1 & -1 \\ -1 & 1 \end{bmatrix} \begin{Bmatrix} T_i \\ T_{i+1} \end{Bmatrix} = \begin{Bmatrix} q_i \\ q_{i+1} \end{Bmatrix}$$

Conductance   Temperature   Heat flow
matrix   vector   vector

or

$$[K]\{T\} = \{q\},$$

where $[K]$ is the element conductance matrix, $\{T\}$ is the vector of nodal temperatures or DOF (degrees of freedom) associated with the heat transfer element, and the vector $\{q\}$ is elemental heat flows. Let us consider the following examples to illustrate the application of heat transfer analysis.

Example 4.5 Consider a typical furnace wall thickness made from steel material, as shown in Fig. 4.10. Model the problem using one element to estimate the temperatures in the furnace wall. The properties are given as: $k = 52$ W/m K; $h_1 = h_2 = 10$ W/m^2/°K; $L = 0.5$ m; $H = 1$ m; $A = 1$ m^2; $T\alpha_1 = 150$°C; and $T\alpha_2 = 25$°C.

Solution: The element conductance matrix, K, is calculated as

$$U_i A \begin{bmatrix} 1 & -1 \\ -1 & 1 \end{bmatrix} \begin{Bmatrix} T_i \\ T_{i+1} \end{Bmatrix} = \begin{Bmatrix} q_i \\ q_{i+1} \end{Bmatrix}.$$

With the given values, we can calculate $U_1 = U_3 = h = 10$; $U_2 = 52/0.5 = 104$; the assembled equations are

$$[K]_G = \begin{bmatrix} U_1 & -U_1 & 0 & 0 \\ -U_1 & U_1 + U_2 & -U_2 & 0 \\ 0 & -U_2 & U_2 + U_3 & -U_3 \\ 0 & 0 & -U_3 & U_3 \end{bmatrix}.$$

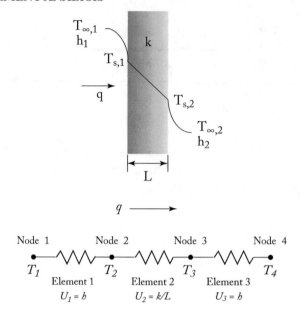

Figure 4.10: Heat transfer through a furnace wall problem (top) and FE model (bottom).

Substituting and applying the boundary conditions $T_1 = 150$ and $T_1 = 25$, we have

$$
\begin{bmatrix}
10 & 0 & 0 & 0 \\
-10 & 114 & -104 & 0 \\
0 & -104 & 114 & -10 \\
0 & 0 & 0 & 10
\end{bmatrix}
\begin{Bmatrix}
150 \\
T_2 \\
T_3 \\
25
\end{Bmatrix}
=
\begin{Bmatrix}
1500 \\
0 \\
0 \\
250
\end{Bmatrix}.
$$

Solving for T_2 and T_3, gives as, $T_2 = 90.4\,°C$ and $T_3 = 84.6\,°C$.

Example 4.6   A typical 1.375 inch solid wood core door is exposed to winter conditions with the characteristics shown in Table 4.1. Assume an inside room temperature of 70°F and an outside air temperature of 20°F, with an exposed area of 22.5 ft². (a) Determine the inside and outside temperature of the door surface, and (b) determine heat loss through the door.

Solution: The finite element model is shown in Fig. 4.11.

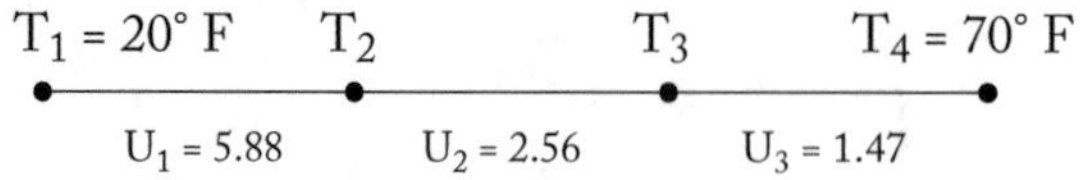

$$T_1 = 20°\ F \qquad T_2 \qquad\qquad T_3 \qquad T_4 = 70°\ F$$

$$U_1 = 5.88 \qquad U_2 = 2.56 \qquad U_3 = 1.47$$

Figure 4.11: The finite element model.

Table 4.1: Characteristics of Example 4.6

Materials/Items	Resistance hr.ft^2.F/Btu	U-factor Btu/hr.ft^2.F
1. Outside file resistance (winter, 15-mph wind)	0.17	5.88
2. 1.375-in solid wood core	0.39	2.56
3. Inside film temperature (winter)	0.68	1.47

(a) Conductance matrix for element #1

$$K^1 = UA \begin{bmatrix} 1 & -1 \\ -1 & 1 \end{bmatrix} = 5.88 \times 22.5 \begin{bmatrix} 1 & -1 \\ -1 & 1 \end{bmatrix}.$$

Conductance matrix for element #2

$$K^2 = UA \begin{bmatrix} 1 & -1 \\ -1 & 1 \end{bmatrix} = 2.56 \times 22.5 \begin{bmatrix} 1 & -1 \\ -1 & 1 \end{bmatrix}.$$

Conductance matrix for element #3

$$K^3 = UA \begin{bmatrix} 1 & -1 \\ -1 & 1 \end{bmatrix} = 1.47 \times 22.5 \begin{bmatrix} 1 & -1 \\ -1 & 1 \end{bmatrix}.$$

Therefore, the global conductance matrix is

$$K^G = 22.5 \times \begin{bmatrix} 5.88 & -5.88 & 0 & 0 \\ -5.88 & 5.88 + 2.56 & -2.56 & 0 \\ 0 & -2.56 & 2.56 + 1.47 & -1.47 \\ 0 & 0 & -1.47 & 1.47 \end{bmatrix}.$$

Apply boundary conditions $T_1 = 20$ F and $T_4 = 70$ F

$$22.5 \times \begin{bmatrix} 5.88 & -5.88 & 0 & 0 \\ -5.88 & 5.88 + 2.56 & -2.56 & 0 \\ 0 & -2.56 & 2.56 + 1.47 & -1.47 \\ 0 & 0 & -1.47 & 1.47 \end{bmatrix} \begin{Bmatrix} 20 \\ T_2 \\ T_3 \\ 70 \end{Bmatrix} = \begin{Bmatrix} q_1 \\ 0 \\ 0 \\ q_4 \end{Bmatrix}.$$

The above system of equations is reduced to

$$22.5 \times \begin{bmatrix} 5.88 + 2.56 & -2.56 \\ -2.56 & 2.56 + 1.47 \end{bmatrix} \begin{Bmatrix} T_2 \\ T_3 \end{Bmatrix} = \begin{Bmatrix} 20 \times 22.5 \times 5.88 \\ 70 \times 22.5 \times 1.47 \end{Bmatrix}.$$

Therefore, $T_2 = 26.85\,°F$ and $T_3 = 42.59\,°F$.

(b) Heat loss through the door is determined as

$$q = 22.5 \times 5.88 \times (26.85 - 20) = 906.55 \text{ Btu/hr.}$$

Example 4.7   The temperature distribution along the fin was approximated using four linear 1D elements. The nodal temperatures and the corresponding positions are shown in Fig. 4.12. Determine the temperature of the fin at the global (X) location (at left support) for (a) $X = 3$ cm and (b) $X = 5.5$ cm using local (x) coordinates.

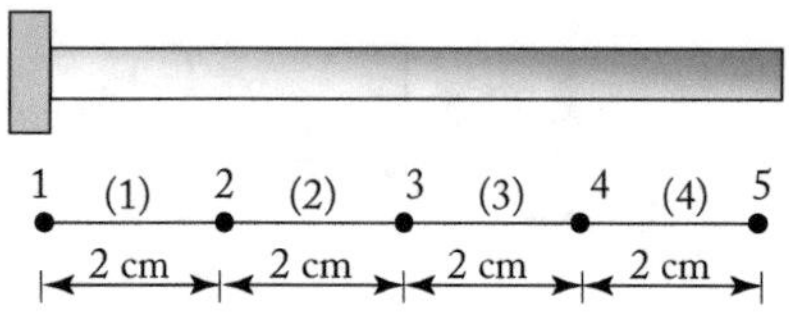

$$\begin{Bmatrix} T_1 \\ T_2 \\ T_3 \\ T_4 \\ T_5 \end{Bmatrix} = \begin{Bmatrix} 100 \\ 75.00 \\ 60.00 \\ 52.0 \\ 48.00 \end{Bmatrix} \text{°C}$$

Figure 4.12: Nodal temperatures and corresponding positions.

Solution:
(a) Temperature at $X = 3$ cm can be described by element #2. Using local coordinates, $x = 1$ cm and $l = 2$ cm

$$T^{(2)} = \left(1 - \frac{x}{l}\right) T_2 + \left(\frac{x}{l}\right) T_3$$

$$T^{(2)} = \left(1 - \frac{1}{2}\right) 75 + \left(\frac{1}{2}\right) 60$$

$$T^{(2)} = 67.5°\text{C.}$$

(b) Temperature at $X = 5.5$ cm can be described by element #3. Using local coordinates, $x = 1.5$ cm and $l = 2$ cm

$$T^{(3)} = \left(1 - \frac{x}{l}\right) T_3 + \left(\frac{x}{l}\right) T_4$$

$$T^{(3)} = \left(1 - \frac{1.5}{2}\right) 60 + \left(\frac{1.5}{2}\right) 52$$

$$T^{(3)} = 46.5°\text{C.}$$

4.6   FLUID ANALYSIS

A fluid is a substance, either gas or liquid, which will deform continuously under the action of applied shear forces on the surface. The discipline of fluid mechanics is concerned with the motion of fluids under different conditions. The governing equations of fluid mechanics are derived from the laws of conservation involving mass, momentum and energy. Conservation of mass gives rise to continuity equations, while the conservation of momentum gives rise to equations of motion, and energy conservation is the first law of thermodynamics. Depending on the type of fluid (air, water, or drug), fluids can be characterized as Newtonian fluids, in which the shear stress is proportional to the rate of deformation with zero stress. The proportionality constant is called the dynamic viscosity, μ. On the other hand, non-Newtonian fluids (colloidal suspensions, emulsions, some polymers) exhibit a variable proportionality between stress and rate of deformation. Also, fluids can be incompressible as in liquids and compressible as in gases and vapors. A fluid can be inviscid or viscous depending on the viscosity of fluid in the analysis. Moreover, depending on the dynamic behavior of the fluid, the motion may be laminar, turbulent, or transitional.

The equation of motion governing the laminar fluid flow inside a channel is given by

$$\mu \frac{d^2 u}{dy^2} - \frac{dP}{dx} = 0,$$

where u is the fluid velocity, μ is the dynamic viscosity of the fluid, dP/dx is the pressure drop across the channel. The FE formulation will result in the following equations:

$$[K] = \frac{\mu}{l} \begin{bmatrix} 1 & -1 \\ -1 & 1 \end{bmatrix}$$

$$\{u\} = \begin{Bmatrix} u_1 \\ u_2 \end{Bmatrix}$$

$$\{F\} = -\frac{\partial P}{\partial x} \frac{l}{2} \begin{Bmatrix} 1 \\ 1 \end{Bmatrix}.$$

The mass flow rate through the channel of width (W) for each fluid element is calculated as

$$massflowrate = \int_{y_i}^{y_j} \rho u W \, dy = \rho W l (u_1 + u_2)/2.$$

For 1D channel problems involving fluid mechanics, the above equations can be used to solve for fluid velocities by specifying the pressure at the two boundaries.

Example 4.8   A drug delivery device that is driven by a pump is shown in Fig. 4.13. The device is used to transport drug from reservoir to outlet. It has a space constraint of 10 mm, hence to accommodate the required length of channel, the channel is curved as shown. Suppose each straight channel is 1 mm long, and each curve has equivalent length of 2 mm. It is assumed that the flow at the reservoir is maintained at 0.1 mm/s and the drug has dynamic viscosity of 1×10^{-2} Pa-s. Determine the flow rate at the outlet, by assuming that outlet is maintained at a small pressure of $P = 1 \times 10^{-3}$ Pa.

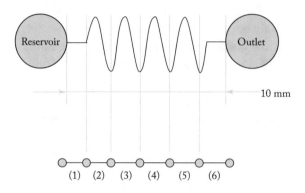

Figure 4.13: Modeling the fluid problem (top) and finite element model (bottom).

Solution: The flow in the channel can be modeled as shown in Fig. 4.13. The reservoir is facilitated with a pump to let the drug flow at the inlet with given speed, v_1, of 0.1 mm/s. It is assumed the turbulence in the bent channel is minimal.

The finite element model shown in Fig. 4.13, has six elements and seven nodes. Using the information given, the element stiffness matrix (after adjusting to the appropriate units) can be constructed as

$$K^1 = \frac{\mu}{l_1} \begin{bmatrix} 1 & -1 \\ -1 & 1 \end{bmatrix} = \frac{1 \times 10^{-2}}{1 \times 10^{-3}} \begin{bmatrix} 1 & -1 \\ -1 & 1 \end{bmatrix}$$

$$K^2 = \frac{\mu}{l_2} \begin{bmatrix} 1 & -1 \\ -1 & 1 \end{bmatrix} = \frac{1 \times 10^{-2}}{2 \times 10^{-3}} \begin{bmatrix} 1 & -1 \\ -1 & 1 \end{bmatrix}.$$

The same procedure can be extended for all other elements. Assembling all the element stiffness matrices, we get the global stiffness matrix as

$$K^G = \begin{bmatrix} 10 & -10 & 0 & 0 & 0 & 0 \\ -10 & 15 & -5 & 0 & 0 & 0 \\ 0 & -5 & 10 & -5 & 0 & 0 \\ 0 & 0 & -5 & 10 & -5 & 0 \\ 0 & 0 & 0 & -5 & 10 & -5 \\ 0 & 0 & 0 & 0 & -10 & 10 \end{bmatrix}.$$

Using the boundary conditions at the inlet of the reservoir ($v_1 = 0.1$ mm/s) and the outlet (pressure of $P = 1 \times 10^{-3}$ Pa), we get the following:

$$\begin{Bmatrix} P_1 \\ P_2 \\ P_3 \\ P_4 \\ P_5 \\ P_6 \\ P_7 \end{Bmatrix} = \begin{Bmatrix} P_1 \\ 0 \\ 0 \\ 0 \\ 0 \\ 0 \\ 1 \times 10^{-3} \end{Bmatrix} \quad \text{and} \quad \begin{Bmatrix} v_1 \\ v_2 \\ v_3 \\ v_4 \\ v_5 \\ v_6 \\ v_7 \end{Bmatrix} = \begin{Bmatrix} 0.1 \times 10^{-3} \\ v_2 \\ v_3 \\ v_4 \\ v_5 \\ v_6 \\ v_7 \end{Bmatrix}.$$

Arranging the equations as

$$\{P\} = \left[K^G\right]\{v\}.$$

Solving the linear system of equations gives velocity of drug at the outlet (node 7) as 0.11 mm/s. If the channel characteristic dimension is 0.01 mm, then the flow rate at outlet is $0.11 * 0.01 = 1.1 \times 10^{-3}$ mm^2/s.

Table 4.2 provides some of the DOF (degrees of freedom) and the corresponding force vectors used in FEA for different engineering problems.

Table 4.2: Summary of FEA in different engineering disciplines

Discipline	DOF	Force Vector
Structural/Solids	Displacement	Mechanical forces
Heat conduction	Temperature	Heat flux
Acoustic fluid	Desplacement potential	Particle velocity
Potential flow	Pressure	Particle velocity
General flows	Velocity	Fluxes
Electrostatics	Electric potential	Charge density
Magnetostatics	Magnetic potential	Magnetic intensity

4.7 FREE VIBRATION ANALYSIS

Many elastic structures having a mass and stiffness, freely vibrate harmonically when disturbed initially at time, $t = 0$. This oscillatory motion is a characteristic property of the structure. Assuming that damping is absent, the oscillatory motion will continue indefinitely, with the amplitude of oscillations depending on the initially imposed disturbance.

Consider a simple spring (k) and mass (m) system with one degree of freedom (x) as shown in Fig. 4.14. The equation of motion by neglecting the damping, can be written as

$$m\ddot{x} + kx = 0.$$

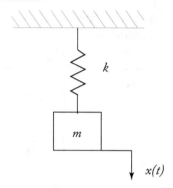

Figure 4.14: Single degree of freedom dynamic system.

The solution to the above equation can be written as

$$x(t) = A \sin \omega t + B \cos \omega t.$$

The natural frequency (ω_n), the time period (T) and the frequency of oscillation (f) are given as

$$\omega_n = \sqrt{\frac{k}{m}}; \qquad T = \frac{2\pi}{\omega}; \qquad f = \frac{1}{T}.$$

The solution to the 1D differential equation can also be written as

$$x = C \sin(\omega t + \phi)$$

$$C = \sqrt{A^2 + B^2} \qquad \phi = \tan^{-1} \frac{B}{A}.$$

The same approach can be extended to multiple dimensions. For free vibration problems, the density of the material is needed which is reflected in the mass matrix. The stiffness and mass matrices along with boundary conditions can be solved to obtain the natural frequencies.

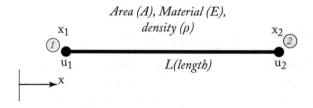

Figure 4.15: Bar element for free vibration analysis.

Consider a two-node bar element shown in Fig. 4.15 with an initial length L, cross-section area A, and density ρ. For free vibration analysis (neglecting the damping in the system), the finite element equations can be written as

$$[M]\{\ddot{u}\} + [K]\{u\} = 0,$$

where $[M]$ is the mass matrix, $[K]$ is the stiffness matrix, and $\{u\}$ is the nodal displacement vector. The mass matrix, also called consistent mass matrix, and stiffness matrix for a two-node bar element is given by

$$[M] = \frac{\rho A L}{6} \begin{bmatrix} 2 & 1 \\ 1 & 2 \end{bmatrix}$$

$$[K] = \frac{EA}{L} \begin{bmatrix} 1 & -1 \\ -1 & 1 \end{bmatrix}.$$

Sometimes, instead of using consistent mass matrix, the lumped mass matrix (in which the mass density is lumped at the end nodes) is used as it simplifies the dynamic equations. The lumped mass matrix for bar element is given by

$$[M] = \frac{\rho A L}{2} \begin{bmatrix} 1 & 0 \\ 0 & 1 \end{bmatrix}.$$

The above equations are demonstrated with a simple example as shown below.

Example 4.9   Consider a uniform bar, which is fixed at the left end and free at the right end as shown in Fig. 4.16. The bar is 1.0 m in length, has a cross-sectional area of 1×10^{-4} m², and is made from steel with Young's modulus of 200 GPa and density of 7850 kg/m³. Find the natural frequency of vibration of the bar using one-element model.

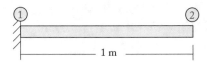

Figure 4.16: Uniform bar modeled as one element.

Solution: Using the stiffness and mass matrices of the bar element, the eigenvalue problem can be written as

$$\frac{EA}{L} \begin{bmatrix} 1 & -1 \\ -1 & 1 \end{bmatrix} \begin{Bmatrix} U_1 \\ U_2 \end{Bmatrix} = \omega^2 \frac{\rho A L}{6} \begin{bmatrix} 2 & 1 \\ 1 & 2 \end{bmatrix} \begin{Bmatrix} U_1 \\ U_2 \end{Bmatrix},$$

where ω is the natural frequency, and U_1 and U_2 are the amplitudes of vibration of the bar at nodes 1 and 2, respectively. By using the boundary condition at the left node ($U_1 = 0$), i.e., we delete the first row and first column in each of the matrices and vectors, and rearranging the resulting equations we get

$$\frac{EA}{L} U_2 = \omega^2 \frac{\rho A L}{6} U_2$$

or

$$\omega = \sqrt{\frac{3E}{\rho L^2}};$$

substituting the given values, we get

$$\omega = \sqrt{\frac{3(2 \times 10^{11})}{7850(1.0)^2}} = 8742.60 \, \text{rad/s}.$$

Similarly, if we use lumped mass matrix instead of consistent mass matrix, we get the natural frequency as

$$\omega = \sqrt{\frac{2E}{\rho L^2}}$$

then substituting the given values, we get

$$\omega = \sqrt{\frac{2(2 \times 10^{11})}{7850(1.0)^2}} = 7138.31 \, \text{rad/s}.$$

The above procedure can be extended to beam and frame elements. The application of free vibration of beams and frames can be found in the bibliography.

4.8　EXERCISE PROBLEMS

4.1. Consider the composite bar made from aluminum ($E = 70$ GPa) and subjected to axial loads as shown in Fig. 4.17. Use the following values: $L_1 = 2.0$ m, $L_2 = 2.5$ m, $P_1 = 4000$ N, $P_2 = 2500$ N, $E_{Al} = 70$ GPa, $A_1 = 20$ cm^2, $A_2 = 10$ cm^2, and determine the displacement of the bar at the applied loads.

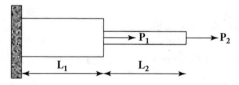

Figure 4.17: Problem 4.1.

4.2. Consider the composite bar consisting of a tapered steel bar fastened to an aluminum rod of uniform cross section and subjected to axial loads as shown in Fig. 4.18. Explain how to set up this problem using the sequence of steps in the FEM and determine the displacement of the bar using FEM. Use the following values: $L_1 = 2.5$ m, $L_2 = 3$ m, $P = 45$ kN, $E_{steel} = 200$ GPa, $E_{Al} = 70$ GPa, $A_{steel}(x = 0) = 15$ cm^2, $A_{Al} = 7.5$ cm^2.

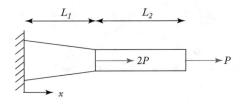

Figure 4.18: Problem 4.2.

4.3. A shaft is made of three parts (AB, BC, and CD). Parts AB and CD are made from the same material ($G = 9.8 \times 10^3$ Ksi) and have a diameter of 1.5 inch. Segment BC is made from a different material ($G = 11.2 \times 10^3$ Ksi) and has a diameter of 1 inch. The shaft is fixed at both ends and subjected to a torque of 1200 lb-in each at B and C. Using three elements, determine the angle of twist at B and C.

4.4. A circular shaft made from two different materials is fixed at points 1 and 3, and subjected to a torque $T = 300$ in-lb at point 2, as shown in Fig. 4.19. Material A is aluminum with a shear modulus $G_A = 4.0 \times 10^6$ lb/in^2, and material B is steel with a shear modulus $G_B = 11.5 \times 10^6$ lb/in^2. Determine the angle of twist at point 2, and the reactions at nodes 1 and 3.

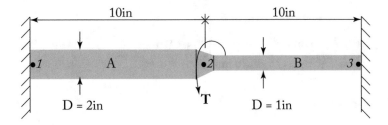

Figure 4.19: Problem 4.4.

4.5. A room in the summer time is maintained at a constant temperature of 23°C. The air outside is at an unpleasant 39°C. Find the temperature distribution inside the wall and the temperature at the inside surface of the pine wood wall, given $k_{\text{pine}} = 0.11$ W/m·k; $k_{\text{insulation1}} = 0.023$ W/m·k; $k_{\text{insulation2}} = 0.035$ W/m·k; and $k_{\text{brick}} = 1.5$ W/m·k (Fig. 4.20).

4.6. A furnace wall (unit area) comprised of a glass fiber insulating material ($k_i = 0.040$ W/m·K) is sandwiched between two steel panels ($k_s = 63.0$ W/m·K). The inner surface of the steel panel is maintained at $T_1 = 200$°C. The surrounding air temperature is $T_\infty = 25$°C, with a convective coefficient of $h = 10$ W/m^2·K. Use FEM to determine the temperature at T_2, T_3, and T_4, and the rate of heat loss through the wall (Fig. 4.21).

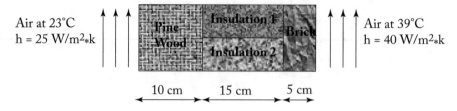

Figure 4.20: Problem 4.5.

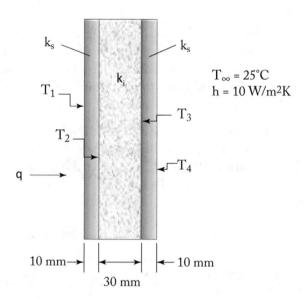

Figure 4.21: Problem 4.6.

4.7. An aluminum plate's inner surface maintained at a constant 200°C against a perfectly insulated wall. The wall next to the plate is shown in Fig. 4.22. The cross sectional area of the plate is 225 cm². Find the temperature distribution within the wall and specifically at the outside surface of the brick wall. Assume the top and bottom to be perfectly insulated. Given $k_{Al} = 0.11$ W/m·k; $k_{insulation1} = 0.08$ W/m·k; and $k_{brick} = 1.5$ W/m·k.

4.8. An oven is heated to bake a pie. The inner air temperature is at a constant 220°C. The oven door is made up of steel ($k_{steel} = 17$ W/m·k), glass ($k_{glass} = 1.05$ W/m·k), and plastic ($k_{plastic} = 0.19$ W/m·k); all of the components are assembled according to the characteristics listed in the Fig. 4.23. The cross-sectional area of the glass portion is 225 cm², the plastic and steel portions both have cross-sectional areas of 1225 cm². Find the temperature distribution within the wall and specifically at the outside surface of the glass and the plastic surfaces. Assume the top and bottom to be perfectly insulated.

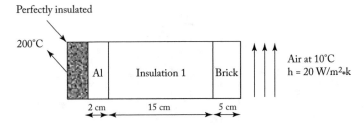

Figure 4.22: Problem 4.7.

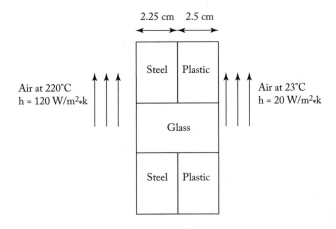

Figure 4.23: Problem 4.8.

4.9. In an effort to reduce heat loss, a pocket of air ($k_{air} = 0.024$ W/m·k) is added to the oven door, as shown in Fig. 4.24. The cross sectional area of the air trapped inside is 1450 cm^2. Find the temperature distribution within the wall and specifically at the outside surface of the glass and the plastic surfaces. Assume the top and bottom to be perfectly insulated.

4.10. Water enters a pipe at a uniform velocity profile and is at 20°C with a viscosity of 1.002 mPa·s and a density of 998.2 kg/m^3. The pipe is shown in Fig. 4.25. Find the velocity at the end of the pipe, given, $L_1 = 2.0$ m, $L_2 = 4$ m, and $d_1 = 50$ mm, $d_2 = 35$ mm. Assume the pressure at the outlet is zero.

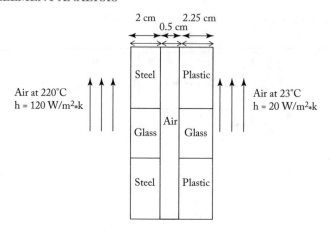

Figure 4.24: Problem 4.9.

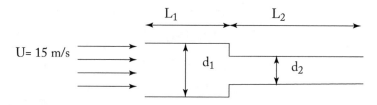

Figure 4.25: Problem 4.10.

4.9   FEA ACTIVITY TEMPLATE

The following template can be used for FEA to systematically solve engineering problems.

1. Conduct a preliminary analysis for a given problem. That is idealize the problem through a free-body diagram, and determine whether any analytical methods can be applied to find an approximate solution.

2. Discretize the model by specifying the geometry, dimensions, nodes, and elements. (Identify suitable elements.)

3. Develop elemental matrices, and assemble them to obtain global matrices.

4. Apply boundary conditions to solve for unknown DOF.

5. Check the solution/results. (Equilibrium, resultant forces, stresses, etc.)

CHAPTER 5

Truss, Beam, and Frame Analysis

After reading this chapter, you will be able to:

- understand the definition of truss and its FE formulation;

- understand the definition of beam and its FE formulation;

- understand the definition of plane frame and its FE formulation;

- understand the definition of space truss; and

- solve truss, beam, and frame problems.

5.1 OVERVIEW

This chapter introduces the FEA tools and techniques for trusses, beams, and frames. Examples are provided for trusses, beams, and frames in order to better illustrate how to find a finite element solution. Based on the examples discussed in this chapter, students should be able to effectively analyze truss, beam, and frame structures.

5.2 TRUSSES

Truss type structures find applications in bridges, power transmission towers, aerospace structural members, building roofs, and others (see Fig. 5.1). A truss is defined as a straight structural member that can support axial loads (tension or compression) and is generally considered as composed of two-force members. Trusses can be defined as plane truss (whose members lie in a plane) or space truss in which the member is in three dimensions. Truss members may consist of metal tubes/bars/channels/angles, and are usually connected through a joint (pin joint in a plane truss or ball-and-socket joint in space truss). Usually trusses with bolted or welded joints can be treated as having smooth pins with little or no bending. The loads acting on the truss are usually applied at the joints and the truss resists the applied loads by developing tension or compression in the members.

Statically determinate trusses are problems that can be solved using the laws of statics. For example, a plane truss problem can be solved using three static equations (summation of

Figure 5.1: Examples of trusses.

forces in two axial directions and one moment equation). However, statically indeterminate truss problems cannot be solved using statics alone. Compatibility equations are often required to solve these problems in addition to statics. Usually, statically indeterminate truss problems are difficult to solve analytically but are very easy to solve using finite element analysis.

5.3   TRUSS FINITE ELEMENT

Consider a truss member/element of length "L" oriented at an angle "θ" with respect to the global X-coordinate as shown in Fig. 5.2. The truss member has a constant cross-sectional area "A," and is made from a material with modulus of elasticity "E." The truss member is subjected to tensile forces "F" along its axis, as shown in Fig. 5.2. Two coordinates systems, a global coordinate system denoted by upper-case X and Y, and a local coordinate system denoted by lower-case x and y are used in the truss analysis.

The truss element has two nodes i and j at each end of the member. With the assumption that the forces on each element act only along the length of the element, thus permitting only axial deformation, the analysis is greatly simplified by attaching a local coordinate system to each member, with the x-axis along its length, as shown in Fig. 5.2.

The nodal forces and displacements in the (local) x-direction are given by f_{ix}, f_{jx} and u_{ix}, u_{jx}, respectively. In local coordinates, there are only forces in the x direction, so $f_{iy} = f_{jy} = 0$ and $u_{iy} = u_{jy} = 0$. The system of equations can then be represented by elemental matrix equation:

$$\{f\} = [K]\{u\}, \quad \text{or,}$$

$$\begin{Bmatrix} f_{ix} \\ f_{iy} \\ f_{jx} \\ f_{jy} \end{Bmatrix} = \begin{bmatrix} k & 0 & -k & 0 \\ 0 & 0 & 0 & 0 \\ -k & 0 & k & 0 \\ 0 & 0 & 0 & 0 \end{bmatrix} \begin{Bmatrix} u_{ix} \\ u_{iy} \\ u_{jx} \\ u_{jy} \end{Bmatrix}. \tag{5.1}$$

The components referred in the local coordinate system can be referred to the global coordinate system by transformation. For instance, the global displacement in the $+X$-direction of node i,

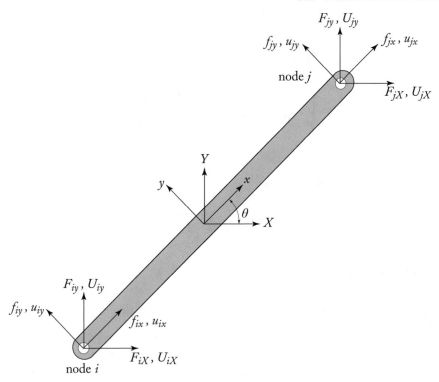

Figure 5.2: A plane truss finite element with global (X, Y) and local (x, y) coordinates.

U_{iX} , can be seen as the sum of the local components in the $+X$-direction, i.e., $U_{iX} = u_{ix} \cos \theta + u_{iy} \cos(\theta + \pi/2) = u_{ix} \cos \theta - u_{iy} \sin \theta$. The global displacements (U) and local displacements (u) are related through the *transformation matrix* $[T]$ which is given as

$$\{U\} = [T]\{u\}, \quad \text{or,}$$

$$\begin{Bmatrix} U_{iX} \\ U_{iY} \\ U_{jX} \\ U_{jY} \end{Bmatrix} = \begin{bmatrix} \cos \theta & -\sin \theta & 0 & 0 \\ \sin \theta & \cos \theta & 0 & 0 \\ 0 & 0 & \cos \theta & -\sin \theta \\ 0 & 0 & \sin \theta & \cos \theta \end{bmatrix} \begin{Bmatrix} u_{ix} \\ u_{iy} \\ u_{jx} \\ u_{jy} \end{Bmatrix}.$$

Similarly, the transformation for the forces is given by:

$$\{F\} = [T]\{f\}, \quad \text{or,}$$

$$\begin{Bmatrix} F_{iX} \\ F_{iY} \\ F_{jX} \\ F_{jY} \end{Bmatrix} = \begin{bmatrix} \cos\theta & -\sin\theta & 0 & 0 \\ \sin\theta & \cos\theta & 0 & 0 \\ 0 & 0 & \cos\theta & -\sin\theta \\ 0 & 0 & \sin\theta & \cos\theta \end{bmatrix} \begin{Bmatrix} f_{ix} \\ f_{iy} \\ f_{jx} \\ f_{jy} \end{Bmatrix}.$$

Rearranging the above equations we get,

$$\{u\} = [T]^{-1}\{U\} \quad \text{and} \quad f = [T]^{-1}\{F\}$$

and substituting into (5.1) yields:

$$\{F\} = [T][K][T]^{-1}\{U\}, \quad \text{or,}$$

$$\begin{Bmatrix} F_{iX} \\ F_{iY} \\ F_{jX} \\ F_{jY} \end{Bmatrix} = k \begin{bmatrix} \cos^2\theta & \sin\theta\cos\theta & -\cos^2\theta & -\sin\theta\cos\theta \\ \sin\theta\cos\theta & \sin^2\theta & -\sin\theta\cos\theta & -\sin^2\theta \\ -\cos^2\theta & -\sin\theta\cos\theta & \cos^2\theta & \sin\theta\cos\theta \\ -\sin\theta\cos\theta & -sin^2\theta & \sin\theta\cos\theta & \sin^2\theta \end{bmatrix} \begin{Bmatrix} U_{iX} \\ U_{iY} \\ U_{jX} \\ U_{jY} \end{Bmatrix}.$$

The first term on the right-hand side of the above equation is the elemental stiffness matrix, which will be designated as $[K]^e$, where e is the element number. The above equations can be written in standard form as

$$\{F\} = [K]\{U\}.$$

Examples of truss analysis are presented next to illustrate the usage of the above equations.

Example 5.1   A solid 304 Stainless Steel truss has a load applied as shown in Fig. 5.3. Each member has a cross sectional area of 27 cm^2 and a Young's modulus of 190 GPa. Find the displacements under the load, reaction forces, and stress in each truss member.

Solution: The truss is modeled using two elements shown in Fig. 5.4.

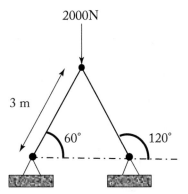

Figure 5.3: Two-bar truss example.

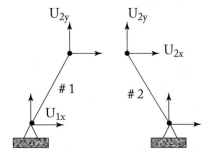

Figure 5.4: Two-element model for the truss.

The truss element stiffness matrix is

$$[K]^e = \frac{EA}{l} \begin{bmatrix} \cos^2\theta & \sin\theta\cos\theta & -\cos^2\theta & -\sin\theta\cos\theta \\ \sin\theta\cos\theta & \sin^2\theta & -\sin\theta\cos\theta & -\sin^2\theta \\ -\cos^2\theta & -\sin\theta\cos\theta & \cos^2\theta & \sin\theta\cos\theta \\ -\sin\theta\cos\theta & -\sin^2\theta & \sin\theta\cos\theta & \sin^2\theta \end{bmatrix}$$

$$[K]^{(1)} = \frac{(190 \times 10^9)(0.0027)}{3}$$
$$\begin{bmatrix} \cos^2 60° & \sin 60°\cos 60° & -\cos^2 60° & -\sin 60°\cos 60° \\ \sin 60°\cos 60° & \sin^2 60° & -\sin 60°\cos 60° & -\sin^2 60° \\ -\cos^2 60° & -\sin 60°\cos 60° & \cos^2 60° & \sin 60°\cos 60° \\ -\sin 60°\cos 60° & -\sin^2 60° & \sin 60°\cos 60° & \sin^2 60° \end{bmatrix}$$

$$[K]^{(1)} = 1.71 \times 10^8 \begin{bmatrix} 0.25 & 0.433 & -0.25 & -0.433 \\ 0.433 & 0.75 & -0.433 & -0.75 \\ -0.25 & -0.433 & 0.25 & 0.433 \\ -0.433 & -0.75 & 0.433 & 0.75 \end{bmatrix}$$

$$[K]^{(2)} = \frac{\left(190 \times 10^9\right)(0.0027)}{3}$$

$$\begin{bmatrix} \cos^2 120° & \sin 120° \cos 120° & -\cos^2 120° & -\sin 120° \cos 120° \\ \sin 120° \cos 120° & \sin^2 120° & -\sin 120° \cos 120° & -\sin^2 120° \\ -\cos^2 120° & -\sin 120° \cos 120° & \cos^2 120° & \sin 120° \cos 120° \\ -\sin 120° \cos 120° & -\sin^2 120° & \sin 120° \cos 120° & \sin^2 120° \end{bmatrix}$$

$$[K]^{(2)} = 1.71 \times 10^8 \begin{bmatrix} 0.25 & -0.433 & -0.25 & 0.433 \\ -0.433 & 0.75 & 0.433 & -0.75 \\ -0.25 & 0.433 & 0.25 & -0.433 \\ 0.433 & -0.75 & -0.433 & 0.75 \end{bmatrix}$$

$$[K]^G = 1.71 \times 10^8 \begin{bmatrix} 0.25 & 0.433 & -0.25 & -0.433 & 0 & 0 \\ 0.433 & 0.75 & -0.433 & -0.75 & 0 & 0 \\ -0.25 & -0.433 & 0.5 & 0 & -0.25 & 0.433 \\ -0.433 & -0.75 & 0 & 1.5 & 0.433 & -0.75 \\ 0 & 0 & -0.25 & 0.433 & 0.25 & -0.433 \\ 0 & 0 & 0.433 & -0.75 & -0.433 & 0.75 \end{bmatrix}$$

$$\{F\}^G = \begin{Bmatrix} 0 \\ 0 \\ 0 \\ -2000 \\ 0 \\ 0 \end{Bmatrix}$$

$$\begin{bmatrix} \left(1.71 \times 10^8\right) 0.5 & 0 \\ 0 & \left(1.71 \times 10^8\right) 1.5 \end{bmatrix} \begin{Bmatrix} U_{2x} \\ U_{2y} \end{Bmatrix} = \begin{Bmatrix} 0 \\ -2000 \end{Bmatrix}$$

$$\begin{Bmatrix} U_{2x} \\ U_{2y} \end{Bmatrix} = \begin{Bmatrix} 0 \\ -7.7973 \times 10^{-6} m \end{Bmatrix}$$

Reaction forces can be calculated as,

$$\{R\} = [K]\{U\} - \{F\}$$

$$\begin{Bmatrix} R_{1x} \\ R_{1y} \\ R_{2x} \\ R_{2y} \\ R_{3x} \\ R_{3y} \end{Bmatrix} = 1.71 \times 10^8 \begin{bmatrix} 0.25 & 0.433 & -0.25 & -0.433 & 0 & 0 \\ 0.433 & 0.75 & -0.433 & -0.75 & 0 & 0 \\ -0.25 & -0.433 & 0.5 & 0 & -0.25 & 0.433 \\ -0.433 & -0.75 & 0 & 1.5 & 0.433 & -0.75 \\ 0 & 0 & -0.25 & 0.433 & 0.25 & -0.433 \\ 0 & 0 & 0.433 & -0.75 & -0.433 & 0.75 \end{bmatrix}$$

$$\begin{Bmatrix} 0 \\ 0 \\ 0 \\ -7.7973 \times 10^{-6} m \\ 0 \\ 0 \end{Bmatrix} - \begin{Bmatrix} 0 \\ 0 \\ 0 \\ -2000 \\ 0 \\ 0 \end{Bmatrix}$$

$$\begin{Bmatrix} R_{1x} \\ R_{1y} \\ R_{2x} \\ R_{2y} \\ R_{3x} \\ R_{3y} \end{Bmatrix} = \begin{Bmatrix} 577.35 \ (N) \\ 1000 \ (N) \\ 0 \\ 0 \\ 577.35 \ (N) \\ 1000 \ (N) \end{Bmatrix}$$

Displacements under the load application can be calculated as,

$$\begin{Bmatrix} u_{2x} \\ u_{2y} \end{Bmatrix} = \begin{bmatrix} \cos 60° & \sin 60° \\ -\sin 60° & \cos 60° \end{bmatrix} \begin{Bmatrix} 0 \\ -7.7973 \times 10^{-6} m \end{Bmatrix}$$

Truss member stress can be calculated as,

$$\begin{Bmatrix} u_{2x} \\ u_{2y} \end{Bmatrix} = \begin{Bmatrix} -6.7527 \times 10^{-6} \\ -3.8987 \times 10^{-6} \end{Bmatrix}$$

$$\sigma^{(1)} = \frac{E(u_{2x} - u_{1x})}{l} = \frac{1.71 \times 10^{-8}(-6.7527 \times 10^{-6} - 0)}{3}$$

$$\sigma^{(1)} = -384.9 \ \text{Pa}$$

$$\sigma^{(2)} = \frac{E(u_{2x} - u_{3x})}{l} = \frac{1.71 \times 10^{-8}(-6.7527 \times 10^{-6} - 0)}{3}$$

$$\sigma^{(2)} = -384.9 \ \text{Pa}.$$

Example 5.2   Consider a four member truss shown in Fig. 5.5. The truss is fixed to the wall and subjected to a force $F = 1000$ lb. Given that all truss members are made from steel (Young's modulus, $E = 30 \times 10^6$ psi) and a cross-sectional area of 1 in^2, find the unknown deflection (DOF).

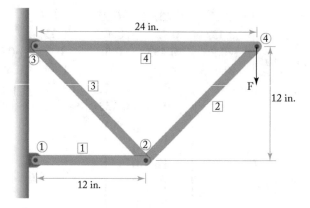

Figure 5.5: Truss example.

Solution: Truss will be modeled using four elements (boxed numbers) with five nodes (circled numbers); as shown in Fig. 5.5.

The stiffness, k, is for element 1 is calculated as:

$$k_1 = \frac{AE}{l_1} = \frac{\left(1 \text{ in}^2\right)\left(30 \times 10^6 \text{ lb/in}^2\right)}{12 \text{ in}} = 2.50 \times 10^6 \text{ lb/in}^2.$$

Similarly, the values of k and for the remaining elements are calculated and summarized in Table 5.1, along with the length of each element and its orientation angle, θ, with respect to the x-axis. Also included are the node numbers for the truss corresponding with the ith and jth nodes of each element.

Table 5.1: Stiffness and orientation of truss elements

Element	Length (in)	Node i	Node j	Stiffness, k (lb/in²) x 10⁶	θ
1	12	1	2	2.50	0°
2	17	2	3	1.67	45°
3	17	2	4	1.67	135°
4	24	3	4	1.25	0°

Starting with element 1, the values for k and θ are used to obtain the local stiffness matrix,

$$[K]_1 = 2.50 \times 10^{-6} \begin{bmatrix} 1 & 0 & -1 & 0 \\ 0 & 0 & 0 & 0 \\ -1 & 0 & 1 & 0 \\ 0 & 0 & 0 & 0 \end{bmatrix} \begin{matrix} U_{1X} \\ U_{1Y} \\ U_{2X} \\ U_{2Y} \end{matrix}.$$

The row labels to the right of the matrix (U_{1X}, U_{2X}, etc.) will be used to locate each of the local stiffness matrices in the global stiffness matrix. The truss has four nodes, each having two degrees of freedom (translation in X and Y directions), which results in an 8×8 global stiffness matrix. The location of $[K]_1$ the global stiffness matrix, denoted as $[K]_{1G}$, is shown below:

$$[K]_{1G} = 10^6 \begin{bmatrix} 2.50 & 0 & -2.50 & 0 & 0 & 0 & 0 & 0 \\ 0 & 0 & 0 & 0 & 0 & 0 & 0 & 0 \\ -2.50 & 0 & 2.50 & 0 & 0 & 0 & 0 & 0 \\ 0 & 0 & 0 & 0 & 0 & 0 & 0 & 0 \\ 0 & 0 & 0 & 0 & 0 & 0 & 0 & 0 \\ 0 & 0 & 0 & 0 & 0 & 0 & 0 & 0 \\ 0 & 0 & 0 & 0 & 0 & 0 & 0 & 0 \\ 0 & 0 & 0 & 0 & 0 & 0 & 0 & 0 \end{bmatrix} \begin{matrix} U_{1X} \\ U_{1Y} \\ U_{2X} \\ U_{2Y} \\ U_{3X} \\ U_{3Y} \\ U_{4X} \\ U_{4Y} \end{matrix}.$$

Element #2 stiffness matrix is developed in the same manner as

$$[K]_2 = 1.76 \times 10^{-6} \begin{bmatrix} 0.5 & 0.5 & -0.5 & -0.5 \\ 0.5 & 0.5 & -0.5 & -0.5 \\ -0.5 & -0.5 & 0.5 & 0.5 \\ -0.5 & -0.5 & 0.5 & 0.5 \end{bmatrix} \begin{matrix} U_{2X} \\ U_{2Y} \\ U_{3X} \\ U_{3Y} \end{matrix}.$$

Its position in the global stiffness matrix is

$$[K]_{2G} = 10^6 \begin{bmatrix} 0 & 0 & 0 & 0 & 0 & 0 & 0 & 0 \\ 0 & 0 & 0 & 0 & 0 & 0 & 0 & 0 \\ 0 & 0 & 0.882 & 0.882 & -0.882 & -0.882 & 0 & 0 \\ 0 & 0 & 0.882 & 0.882 & -0.882 & -0.882 & 0 & 0 \\ 0 & 0 & -0.882 & -0.882 & 0.882 & 0.882 & 0 & 0 \\ 0 & 0 & -0.882 & -0.882 & 0.882 & 0.882 & 0 & 0 \\ 0 & 0 & 0 & 0 & 0 & 0 & 0 & 0 \\ 0 & 0 & 0 & 0 & 0 & 0 & 0 & 0 \end{bmatrix} \begin{matrix} U_{1X} \\ U_{1Y} \\ U_{2X} \\ U_{2Y} \\ U_{3X} \\ U_{3Y} \\ U_{4X} \\ U_{4Y} \end{matrix}.$$

Continuing the process for the remaining elements and adding the matrices results in the final global matrix as

$$K_G = K_1 + K_2 + K_3 + K_4$$

$$[K]_G = 10^6 \begin{bmatrix} 2.50 & 0 & -2.50 & 0 & 0 & 0 & 0 & 0 \\ 0 & 0 & 0 & 0 & 0 & 0 & 0 & 0 \\ -2.50 & 0 & 4.26 & 0 & -0.882 & -0.882 & -0.882 & 0.882 \\ 0 & 0 & 0 & 1.76 & -0.882 & -0.882 & 0.882 & -0.882 \\ 0 & 0 & -0.882 & -0.882 & 2.13 & 0.833 & -1.25 & 0 \\ 0 & 0 & -0.882 & -0.882 & 0.882 & 0.882 & 0 & 0 \\ 0 & 0 & -0.882 & 0.882 & -1.25 & 0 & 2.13 & -0.882 \\ 0 & 0 & 0.882 & -0.882 & 0 & 0 & -0.882 & 0.882 \end{bmatrix}.$$

Substituting the global stiffness matrix and force and displacement vectors into the matrix equation gives

$$10^6 \begin{bmatrix} 2.50 & 0 & -2.50 & 0 & 0 & 0 & 0 & 0 \\ 0 & 0 & 0 & 0 & 0 & 0 & 0 & 0 \\ -2.50 & 0 & 4.26 & 0 & -0.882 & -0.882 & -0.882 & 0.882 \\ 0 & 0 & 0 & 1.76 & -0.882 & -0.882 & 0.882 & -0.882 \\ 0 & 0 & -0.882 & -0.882 & 2.13 & 0.833 & -1.25 & 0 \\ 0 & 0 & -0.882 & -0.882 & 0.882 & 0.882 & 0 & 0 \\ 0 & 0 & -0.882 & 0.882 & -1.25 & 0 & 2.13 & -0.882 \\ 0 & 0 & 0.882 & -0.882 & 0 & 0 & -0.882 & 0.882 \end{bmatrix} \begin{Bmatrix} U_{1X} \\ U_{1Y} \\ U_{2X} \\ U_{2Y} \\ U_{3X} \\ U_{3Y} \\ U_{4X} \\ U_{4Y} \end{Bmatrix}$$

$$= \begin{Bmatrix} F_{1X} \\ F_{1Y} \\ F_{2X} \\ F_{2Y} \\ F_{3X} \\ F_{3Y} \\ F_{4X} \\ F_{4Y} \end{Bmatrix}.$$

Recognizing that nodes 1 and 3 are fixed, such that U_{1X}, U_{1Y}, U_{3X}, and U_{3Y} all equal zero, their corresponding rows and columns in the matrices (rows and columns 1, 2, 5, and 6) can be eliminated, and substituting the applied force at node 4 for F_{4Y}, leaves the following equation to be solved:

$$10^6 \begin{bmatrix} 4.26 & 0 & -0.882 & 0.882 \\ 0 & 1.76 & 0.882 & -0.882 \\ -0.882 & 0.882 & 2.13 & -0.882 \\ 0.882 & -0.882 & -0.882 & 0.882 \end{bmatrix} \begin{Bmatrix} U_{2X} \\ U_{2Y} \\ U_{4X} \\ U_{4Y} \end{Bmatrix} = \begin{Bmatrix} 0 \\ 0 \\ 0 \\ -1000 \end{Bmatrix}.$$

Solving the matrix equation for unknowns in U we get:

$$\{U\} = \left\{ \begin{array}{c} -0.800 \\ -1.93 \\ 0.800 \\ -4.70 \end{array} \right\} \times 10^{-3} \text{ in.}$$

5.4 SPACE TRUSSES

A space truss is a 3D structure, and has three DOFs, u, v, and w, at each node of the truss. Space truss type structures find in many engineering applications, including space satellite members, power line cable structures, and steam energy generation components. An example of 3D truss is shown in Fig. 5.6. In the analysis of space trusses, the joints are assumed to be ball-and-socket joints with negligible bending moments. Also, all the loads are applied at the joints.

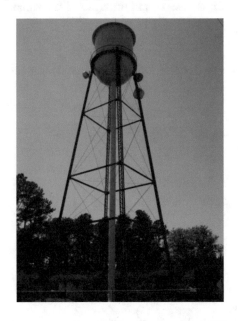
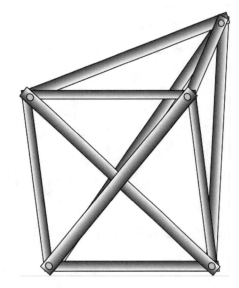

Figure 5.6: Examples of space truss (3D).

The stiffness matrix of space trusses is an extension of and is similar to the plane truss stiffness matrix. At each node of the space truss, there are three global displacements representing the movement in three directions. Therefore, the nodal vector for space truss element will have 6 DOF ($U_{iX}, U_{iY}, U_{iZ}, U_{jX}, U_{jY}, U_{jZ}$). The orientation of a space truss member is defined by the angles θ_X, θ_Y, and θ_Z, with respect to the global coordinate system (X, Y, Z), as shown in Fig. 5.7.

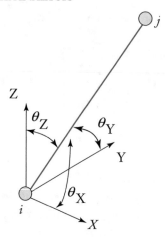

Figure 5.7: Space truss member orientation defined by direction cosines and w.r.t. global coordinate system $(X, Y, \text{and } Z)$.

The elemental stiffness matrix for a space truss s of size 6×6 (6 DOF per element) is given by

$$[\bar{K}]^{(e)} = K \begin{bmatrix} c_X^2 & c_X c_Y & c_X c_Z & -c_X^2 & -c_X c_Y & c_X c_Z \\ c_X c_Y & c_Y^2 & c_Y c_Z & -c_X c_Y & -c_Y^2 & -c_Y c_Z \\ c_X c_Z & c_Y c_Z & c_Z^2 & -c_X c_Z & -c_Y c_Z & -c_Z^2 \\ -c_X^2 & -c_X c_Y & -c_X c_Z & c_X^2 & c_X c_Y & c_X c_Z \\ -c_X c_Y & -c_Y^2 & -c_Y c_Z & c_X c_Y & c_Y^2 & c_Y c_Z \\ -c_X c_Z & -c_Y c_Z & -c_Z^2 & c_X c_Z & c_Y c_Z & c_Z^2 \end{bmatrix}$$

where $k = (EA/L)$ and

$$c_X = \cos \theta_X, \quad c_Y = \cos \theta_Y, \quad c_Z = \cos \theta_Z$$

$$\cos \theta_X = \frac{X_j - X_i}{L}, \quad \cos \theta_Y = \frac{Y_j - Y_i}{L}, \quad \cos \theta_Z = \frac{Z_j - Z_i}{L}$$

$$L = \sqrt{(X_j - X_i)^2 + (Y_j - Y_i)^2 + (Z_j - Z_i)^2}.$$

The process of assembly, applying loads and boundary conditions, and solving for unknown DOF is exactly the same as that followed in solving a 2D truss problem. An example of 3D truss analysis using ANSYS is discussed in later chapters.

5.5   BEAMS

A beam is defined as a structural member that can support transverse loads which will cause bending. Usually, the length of the beam is much longer than the cross-sectional dimensions. Beam-type structures are found in many engineering applications, including aircraft structures, buildings, bridges, automotive and energy components. Figure 5.8 shows examples of beam-type structures.

Figure 5.8: Examples of beam-type structures.

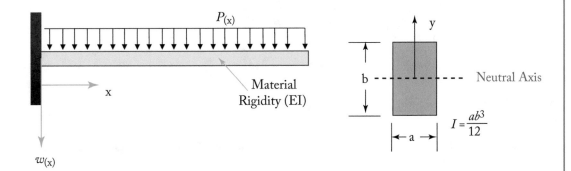

Figure 5.9: A typical beam with loading.

Consider a beam subjected to loading, as shown in Fig. 5.9. The governing differential equation, bending moment and the bending stress are given below:

$$\text{Differential Equation:} \quad EI\frac{d^4w}{dx^4} = P(x).$$

$$\text{Transverse Shear:} \quad EI\frac{d^3w}{dx^3} = V(x).$$

$$\text{Bending Moment:} \quad EI\frac{d^2w}{dx^2} = M(x).$$

The flexural formula, which relates bending stress (σ) to moment at any section $(M(x))$, and moment of inertia (I) is given by

$$\sigma = \frac{M(x)y}{I},$$

where y is the distance from the neutral axis.

The governing differential equation is solved using the two boundary conditions each at the left and right supports. The boundary conditions include fixed (both displacement and rotation is zero); pinned/hinged (displacement is zero and moment is zero); and free (both shear and moment are zero).

5.6   BEAM FINITE ELEMENT EQUATIONS

Consider a beam element of length "L" with a constant moment of inertia "I," made from a material with modulus of elasticity "E," as shown in Fig. 5.10.

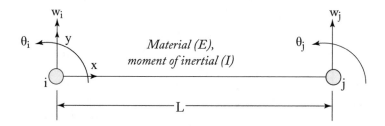

Figure 5.10: Beam finite element.

The beam element has two nodes with nodal displacements as w_i, θ_i, and w_j, θ_j, at nodes i and j, respectively.

Assume that the vertical displacement (w) within a beam element is given by

$$w^e(x) = c_1 + c_2 x + c_3 x^2 + c_4 x^3.$$

The four unknown constants $(c_1, c_2, c_3,$ and $c_4)$ are found by using the end conditions as

$$x = 0, \; w = w_i$$
$$x = 0, \; \frac{\partial w}{\partial x} = \theta_i$$
$$x = L, \; w = w_j$$
$$x = L, \; \frac{\partial w}{\partial x} = \theta_j.$$

The deflection (w) can be also written in terms of shape functions in local coordinate (x) as

$$w^e(x) = \{ \begin{array}{cccc} S_{i1} & S_{i2} & S_{j1} & S_{j2} \end{array} \} \left\{ \begin{array}{c} w_i \\ \theta_i \\ w_j \\ \theta_j \end{array} \right\}$$

where

$$S_{i1} = 1 - \frac{3x^2}{L^2} + \frac{2x^3}{L^3}$$

$$S_{i2} = x - \frac{2x^2}{L} + \frac{x^3}{L^2}$$

$$S_{j1} = \frac{3x^2}{L^2} - \frac{2x^3}{L^3}$$

$$S_{j2} = -\frac{x^2}{L} + \frac{x^3}{L^2}.$$

To find the stiffness matrix, using the total minimum potential energy method,

$$\frac{d}{dx} \left(\frac{1}{2} \sigma \varepsilon + WorkDone \right) = 0 \Rightarrow [K]$$

$$[K]^e = \int_x [B]^T [C] [B] \, dx,$$

where $[B]$ is the strain displacement matrix and $[C]$ is the constitutive matrix which relates axial stress (σ) to bending strain (ε) through Hooke's law. After simplifying, the beam element stiffness matrix is given by

$$[K]^e = \frac{EI}{L^3} \begin{bmatrix} 12 & 6L & -12 & 6L \\ 6L & 4L^2 & -6L & 2L^2 \\ -12 & -6L & 12 & -6L \\ 6L & 2L^2 & -6L & 4L^2 \end{bmatrix}.$$

The finite element formulation for 1D beam element results in the equations of the form as

$$[K] = \{u\} = \{f\},$$

where $[K]$ is the beam element stiffness matrix, $\{u\}$ is the vector of nodal DOF associated with the beam element, and the vector $\{f\}$ is the nodal forces. The load matrix for a beam element can be found using equivalent beam nodal reactions corresponding to a given loading. For typical loading cases, the equivalent nodal loads are given in Fig. 5.11 below.

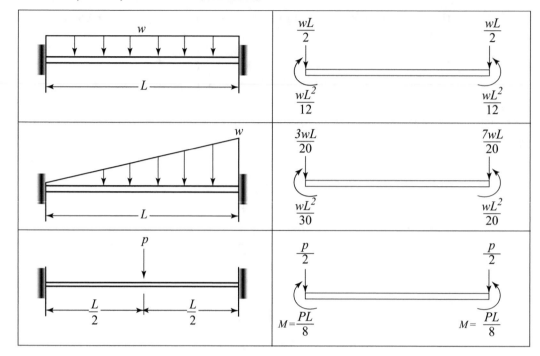

Figure 5.11: Equivalent nodal loads for typical loading on beams.

Let's consider the following example to illustrate the application of beam element for stress analysis.

Example 5.3   A steel cantilever beam $(E = 190 \text{ GPa})$ is subjected to a uniformly distributed load as shown in Fig. 5.12. The beam has a width of 15 cm and a height of 30 cm. Find the unknown degrees of freedom using two element solution.

Solution: The finite element model for the beam with two elements and 3 nodes is shown in Fig. 5.13.

Pre-processing:
    Second moment of area is given as

$$I = \frac{1}{12}bh^3 = \left(\frac{1}{12}\right)(30 \times 10^{-2} \text{ m})(15 \times 10^{-2} \text{ m})^3$$
$$I = 8.4375 \times 10^{-5} \text{ m}^4.$$

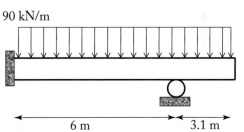

Figure 5.12: Beam example.

Figure 5.13: Finite element model.

Element nodal forces using Fig. 5.11 and are calculated as

$$\{F\}^{(1)} = \left\{\begin{array}{c} f_{a1} \\ f_{a2} \\ f_{b1,1} \\ f_{b2,1} \end{array}\right\} = \left\{\begin{array}{c} -\frac{wL}{2} \\ -\frac{wL^2}{12} \\ -\frac{wL}{2} \\ \frac{wL^2}{12} \end{array}\right\} = \left\{\begin{array}{c} -\frac{90000\times6}{2} \\ -\frac{90000\times6^2}{12} \\ -\frac{90000\times6}{2} \\ \frac{90000\times6^2}{12} \end{array}\right\}$$

$$\{F\}^{(1)} = \left\{\begin{array}{c} -270000 \\ -270000 \\ -270000 \\ 270000 \end{array}\right\}$$

$$\{F\}^{(2)} = \left\{\begin{array}{c} f_{b1,2} \\ f_{b2,2} \\ f_{c1} \\ f_{c2} \end{array}\right\} = \left\{\begin{array}{c} -\frac{wL}{2} \\ -\frac{wL^2}{12} \\ -\frac{wL}{2} \\ \frac{wL^2}{12} \end{array}\right\} = \left\{\begin{array}{c} -\frac{90000\times3.1}{2} \\ -\frac{90000\times3.1^2}{12} \\ -\frac{90000\times3.1}{2} \\ \frac{90000\times3.1^2}{12} \end{array}\right\}$$

$$\{F\}^{(2)} = \begin{Bmatrix} -139500 \\ -72075 \\ -139500 \\ 72075 \end{Bmatrix}$$

$$\{F\}^G = \begin{Bmatrix} f_{a1} \\ f_{a2} \\ f_{b1,1} + f_{b1,2} \\ f_{b2,1} + f_{b2,2} \\ f_{c1} \\ f_{c2} \end{Bmatrix} = \begin{Bmatrix} -270000 \\ -270000 \\ -270000 - 139500 \\ 270000 - 72075 \\ -139500 \\ 72075 \end{Bmatrix} = \begin{Bmatrix} -270000 \\ -270000 \\ -409500 \\ 197925 \\ -139500 \\ 72075 \end{Bmatrix}.$$

Stiffness matrix

$$[K]^{(e)} = \frac{EI}{L^3} \begin{bmatrix} 12 & 6L & -12 & 6L \\ 6L & 4L^2 & -6L & 2L^2 \\ -12 & -6L & 12 & -6L \\ 6L & 2L^2 & -6L & 4L^2 \end{bmatrix}$$

$$[K]^{(1)} = \frac{\left(190 \times 10^9\right)\left(8.4375 \times 10^{-5}\right)}{6^3} \begin{bmatrix} 12 & 6\,(6) & -12 & 6\,(6) \\ 6\,(6) & 4(6)^2 & -6\,(6) & 2(6)^2 \\ -12 & -6\,(6) & 12 & -6\,(6) \\ 6\,(6) & 2(6)^2 & -6\,(6) & 4(6)^2 \end{bmatrix}$$

$$[K]^{(1)} = 890625 \begin{bmatrix} 1 & 3 & -1 & 3 \\ 3 & 12 & -3 & 6 \\ -1 & -3 & 1 & -3 \\ 3 & 6 & -3 & 12 \end{bmatrix}$$

$$[K]^{(2)} = \frac{\left(190 \times 10^9\right)\left(8.4375 \times 10^{-5}\right)}{3.1^3} \begin{bmatrix} 12 & 6\,(3.1) & -12 & 6\,(3.1) \\ 6\,(3.1) & 4(3.1)^2 & -6\,(3.1) & 2(3.1)^2 \\ -12 & -6\,(3.1) & 12 & -6\,(3.1) \\ 6\,(3.1) & 2(3.1)^2 & -6\,(3.1) & 4(3.1)^2 \end{bmatrix}$$

$$[K]^{(2)} = 538123.93 \begin{bmatrix} 12 & 18.6 & -12 & 18.6 \\ 18.6 & 38.44 & -18.6 & 19.22 \\ -12 & -18.6 & 12 & -18.6 \\ 18.6 & 19.22 & -18.6 & 38.44 \end{bmatrix}$$

$[K]^G =$

$$
\begin{bmatrix}
890625 & 2671875 & -890625 & 2671875 & 0 & 0 \\
2671875 & 10687500 & -2671875 & 5343750 & 0 & 0 \\
-890625 & -2671875 & (890625 + 6457487.16) & (-2671875 + 10009105.1) & -6457487.16 & 10009105.1 \\
2671875 & 5343750 & (-2671875 + 10009105.1) & (10687500 + 20685483.87) & -10009105.1 & 10342741.93 \\
0 & 0 & -6457487.16 & -10009105.1 & 6457487.16 & -10009105.1 \\
0 & 0 & 10009105.1 & 10342741.93 & -10009105.1 & 20685483.87
\end{bmatrix}
$$

$$
[K]^G =
\begin{bmatrix}
890625 & 2671875 & -890625 & 2671875 & 0 & 0 \\
2671875 & 10687500 & -2671875 & 5343750 & 0 & 0 \\
-890625 & -2671875 & 7348112.16 & 7337230.1 & -6457487.16 & 10009105.1 \\
2671875 & 5343750 & 7337230.1 & 31372983.87 & -10009105.1 & 10342741.93 \\
0 & 0 & -6457487.16 & -10009105.1 & 6457487.16 & -10009105.1 \\
0 & 0 & 10009105.1 & 10342741.93 & -10009105.1 & 20685483.87
\end{bmatrix} .
$$

Solution to find displacements

$$[\boldsymbol{K}]\{\boldsymbol{U}\} = \{\boldsymbol{F}\}$$

$$
\begin{bmatrix}
1 & 0 & 0 & 0 & 0 & 0 \\
0 & 1 & 0 & 0 & 0 & 0 \\
0 & 0 & 1 & 0 & 0 & 0 \\
2671875 & 5343750 & 7337230.1 & 31372983.87 & -10009105.1 & 10342741.93 \\
0 & 0 & -6457487.16 & -10009105.1 & 6457487.16 & -10009105.1 \\
0 & 0 & 10009105.1 & 10342741.93 & -10009105.1 & 20685483.87
\end{bmatrix}
$$

$$
\begin{Bmatrix}
U_{a1} \\
U_{a2} \\
U_{b1} \\
U_{b2} \\
U_{c1} \\
U_{c2}
\end{Bmatrix}
=
\begin{Bmatrix}
0 \\
0 \\
0 \\
197925 \\
-139500 \\
72075
\end{Bmatrix}
$$

$$
\begin{Bmatrix}
U_{a1} \\
U_{a2} \\
U_{b1} \\
U_{b2} \\
U_{c1} \\
U_{c2}
\end{Bmatrix}
=
\begin{Bmatrix}
0 \\
0 \\
0 \\
-0.015187 \text{ rad} \\
-0.111859 \text{ m} \\
-0.043048 \text{ rad}
\end{Bmatrix} .
$$

Solution to find reactions

$$[\boldsymbol{R}] = [\boldsymbol{K}]\{\boldsymbol{U}\} - \{\boldsymbol{F}\}$$

$$
\begin{Bmatrix} R_a \\ M_a \\ R_b \\ M_b \\ R_c \\ M_c \end{Bmatrix} =
$$

$$
\begin{bmatrix}
890625 & 2671875 & -890625 & 2671875 & 0 & 0 \\
2671875 & 10687500 & -2671875 & 5343750 & 0 & 0 \\
-890625 & -2671875 & 7348112.16 & 7337230.1 & -6457487.16 & 10009105.1 \\
2671875 & 5343750 & 7337230.1 & 31372983.87 & -10009105.1 & 10342741.93 \\
0 & 0 & -6457487.16 & -10009105.1 & 6457487.16 & -10009105.1 \\
0 & 0 & 10009105.1 & 10342741.93 & -10009105.1 & 20685483.87
\end{bmatrix}
$$

$$
\begin{Bmatrix} 0 \\ 0 \\ 0 \\ -0.015187 \text{ rad} \\ -0.111859 \text{ m} \\ -0.043048 \text{ rad} \end{Bmatrix}
-
\begin{Bmatrix} -270000 \\ -270000 \\ -409500 \\ 197925 \\ -139500 \\ 72075 \end{Bmatrix}
$$

$$
\begin{Bmatrix} R_a \\ M_a \\ R_b \\ M_b \\ R_c \\ M_c \end{Bmatrix} =
\begin{Bmatrix} 229422.8 \text{ (N)} \\ 188845.7 \text{ (Nm)} \\ 589531.5 \text{ (N)} \\ 0 \\ 0 \\ 0 \end{Bmatrix}.
$$

5.7   FRAME ELEMENT FORMULATION

A frame is a structural member similar to a truss but can support both axial and bending loads. A frame member combines both an axial bar and a beam by connecting the members rigidly with welded or bolted joints through a gusset plate. Frame elements unlike beam elements, can support both axial loads as well as transverse loads. Consider a plane frame, which can move only in a two-dimensional plane. Figure 5.14 shows a frame element, which consists of two nodes, and at each node, there are three degrees of freedom—a longitudinal displacement, a lateral displacement, and a rotation.

As can be seen from Fig. 5.14, at node i, u_{i1} represents the longitudinal displacement, u_{i2} represents the lateral displacement and u_{i3} represents the rotation. Similarly, at node j, u_{j1}, u_{j2},

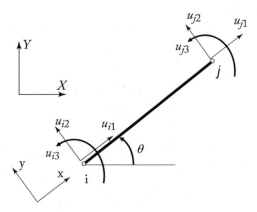

Figure 5.14: A frame element (with local coordinates (x, y), global coordinates (X, Y), and local degrees of freedom).

and u_{j3} represent the longitudinal displacement, the lateral displacement and the rotation, respectively. Two coordinate systems are used to describe the frame elements: a global coordinate system (X, Y) and a local coordinate system (x, y). The frame element in local coordinates (x, y) is given as

$$[K]^{(e)}\{u\} = \{f\}^{(e)},$$

where

$$
\frac{E}{L}
\begin{bmatrix}
A & 0 & 0 & -A & 0 & 0 \\
0 & 12l/L^2 & 6l/L & 0 & -12l/L^2 & 6l/L \\
0 & 6l/L & 4l & 0 & -6l/L & 2l \\
-A & 0 & 0 & A & 0 & 0 \\
0 & -12l/L^2 & -6l/L & 0 & 12l/L^2 & -6l/L \\
0 & 6l/L & 2l & 0 & -6l/L & 4l
\end{bmatrix}
\begin{Bmatrix}
u_{i1} \\
u_{i2} \\
u_{i3} \\
u_{j1} \\
u_{j2} \\
u_{j3}
\end{Bmatrix}
= \{f\}^{(e)}.
$$

In the global coordinate system, which is fixed, used to represent orientation (θ) of each frame element, apply loads and constraints, and to represent the solution. The local coordinate system will be used to represent each element's behavior. The relationship between the local coordinate system (x, y) and the global coordinate system (X, Y) is given through a transformation matrix following Fig. 5.14 as

$$
[T] =
\begin{bmatrix}
c & s & 0 & 0 & 0 & 0 \\
-s & c & 0 & 0 & 0 & 0 \\
0 & 0 & 1 & 0 & 0 & 0 \\
0 & 0 & 0 & c & s & 0 \\
0 & 0 & 0 & -s & c & 0 \\
0 & 0 & 0 & 0 & 0 & 1
\end{bmatrix},
$$

where $c = \cos\theta$ and $s = \sin\theta$.

The stiffness matrix in the global coordinate system (X, Y) is given as

$$
\begin{aligned}
[\bar{\boldsymbol{K}}]^{(e)}\{\boldsymbol{U}\} &= \{\boldsymbol{F}\}^{(e)} \\
[\bar{\boldsymbol{K}}]^{(e)} &= [\boldsymbol{T}]^{-1}[\boldsymbol{K}][\boldsymbol{T}] \\
\{\boldsymbol{F}\}^{(e)} &= [\boldsymbol{T}]^{-1}\{\boldsymbol{f}\}^{(e)}.
\end{aligned}
$$

For a uniformly loaded frame element as shown in Fig. 5.15, the load vector is given as

$$
\{\boldsymbol{f}\}^{(e)} = \begin{Bmatrix} fL/2 \\ wL/2 \\ wL^2/12 \\ fL/2 \\ wL/2 \\ -wL^2/12 \end{Bmatrix} + \begin{Bmatrix} R_{i1} \\ R_{i2} \\ R_{i3} \\ R_{j1} \\ R_{j2} \\ R_{j3} \end{Bmatrix}.
$$

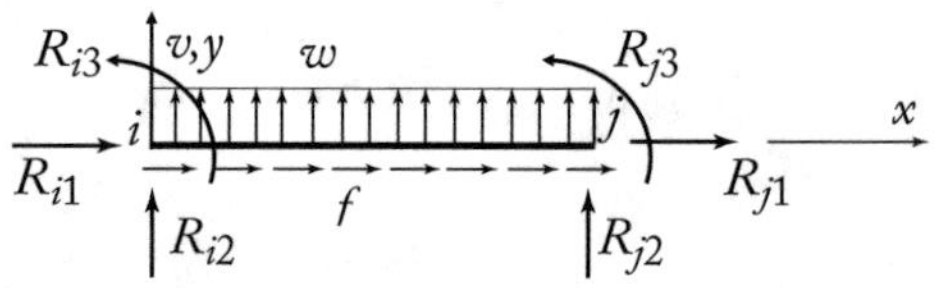

Figure 5.15: A frame element with uniform loading.

Example 5.4 Consider the frame made from a material ($E = 30 \times 10^6$ lb/in^2) and the properties shown in Fig. 5.16. Model the frame using two elements and find unknown displacements.

Solution:

Stiffness matrices:

$$
[\boldsymbol{K}]^e_{xy} = \begin{bmatrix}
\frac{AE}{L} & 0 & 0 & -\frac{AE}{L} & 0 & 0 \\
0 & \frac{12EI}{L^3} & \frac{6EI}{L^2} & 2L^2 & -\frac{12EI}{L^3} & \frac{6EI}{L^2} \\
0 & \frac{6EI}{L^2} & \frac{4EI}{L} & -6L & -\frac{6EI}{L^2} & \frac{2EI}{L} \\
6L & 2L^2 & -6L & \frac{AE}{L} & 0 & 0 \\
o & -\frac{12EI}{L^3} & -\frac{6EI}{L^2} & 0 & \frac{12EI}{L^3} & -\frac{6EI}{L^2} \\
o & \frac{6EI}{L^2} & \frac{2EI}{L} & 0 & -\frac{6EI}{L^2} & \frac{4EI}{L}
\end{bmatrix}.
$$

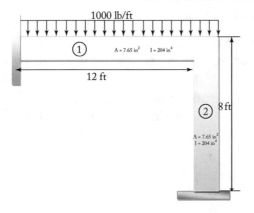

Figure 5.16: Frame example.

For element (1)

$$[K]^{(1)} = 10^3 \begin{bmatrix} 1593.75 & 0 & 0 & -1593.75 & 0 & 0 \\ 0 & 24.595 & 1770.833 & 0 & -24.595 & 1770.833 \\ 0 & 1770.833 & 170000 & 0 & -1770.83 & 85000 \\ -1593.75 & 0 & 0 & 1593.75 & 0 & 0 \\ 0 & -24.595 & -1770.83 & 0 & 24.595 & -1770.83 \\ 0 & 1770.833 & 85000 & 0 & -1770.83 & 170000 \end{bmatrix}.$$

For element (2), the stiffness matrix represented with respect to the local coordinate system is

$$[K]_{xy}^{(2)} = 10^3 \begin{bmatrix} 2390.625 & 0 & 0 & -2390.625 & 0 & 0 \\ 0 & 83.008 & 3984.375 & 0 & -83.008 & 3984.375 \\ 0 & 3984.375 & 255000 & 0 & -3984.375 & 127500 \\ -2390.625 & 0 & 0 & 2390.625 & 0 & 0 \\ 0 & -83.008 & -3984.375 & 0 & 83.008 & -3984.375 \\ 0 & 3984.375 & 127500 & 0 & -3984.375 & 255000 \end{bmatrix}.$$

For element #2, the transformation matrix is

$$[T] = \begin{bmatrix} \cos(270) & \sin(270) & 0 & 0 & 0 & 0 \\ -\sin(270) & \cos(270) & 0 & 0 & 0 & 0 \\ 0 & 0 & 1 & 0 & 0 & 0 \\ 0 & 0 & 0 & \cos(270) & \sin(270) & 0 \\ 0 & 0 & 0 & -\sin(270) & \cos(270) & 0 \\ 0 & 0 & 0 & 0 & 0 & 1 \end{bmatrix} = \begin{bmatrix} 0 & -1 & 0 & 0 & 0 & 0 \\ 1 & 0 & 0 & 0 & 0 & 0 \\ 0 & 0 & 1 & 0 & 0 & 0 \\ 0 & 0 & 0 & 0 & -1 & 0 \\ 0 & 0 & 0 & 1 & 0 & 0 \\ 0 & 0 & 0 & 0 & 0 & 1 \end{bmatrix}$$

$$[T]^T = \begin{bmatrix} 0 & -1 & 0 & 0 & 0 & 0 \\ 1 & 0 & 0 & 0 & 0 & 0 \\ 0 & 0 & 1 & 0 & 0 & 0 \\ 0 & 0 & 0 & 0 & 1 & 0 \\ 0 & 0 & 0 & -1 & 0 & 0 \\ 0 & 0 & 0 & 0 & 0 & 1 \end{bmatrix}.$$

Therefore,

$$[K]^2 = [T]^T [K]^{(2)}_{xy} [T]$$

$$= \begin{bmatrix} 83.008 & 0 & 3984.375 & -83.008 & 0 & 3984.375 \\ 0 & 2390.625 & 0 & 0 & -2390.625 & 0 \\ 3984.375 & 0 & 255000 & -3984.375 & 0 & 127500 \\ -83.008 & 0 & -3984.375 & 83.008 & 0 & -3984.375 \\ 0 & -2390.625 & 0 & 0 & 2390.625 & 0 \\ 3984.375 & 0 & 127500 & -3984.375 & 0 & 255000 \end{bmatrix}.$$

Constructing the global stiffness matrix by assembling $[K]^{(1)}$ and $[K]^{(2)}$, we have

$$[K] = 10^3 \begin{bmatrix} 1593.75 & 0 & 0 & -1593.75 & 0 \\ 0 & 24.595 & 1770.833 & 0 & -24.595 \\ 0 & 1770.833 & 170000 & 0 & -1770.83 \\ -1593.75 & 0 & 0 & 1593.75 + 83.008 & 0 \\ 0 & -24.595 & -1770.83 & 0 & 24.595 + 2390.625 \\ 0 & 1770.833 & 85000 & 0 + 3984.375 & -1770.83 \\ 0 & 0 & 0 & -83.008 & 0 \\ 0 & 0 & 0 & 0 & -2390.625 \\ 0 & 0 & 0 & 3984.375 & 0 \end{bmatrix}$$

$$\begin{bmatrix} 0 & 0 & 0 & 0 \\ 770.833 & 0 & 0 & 0 \\ 85000 & 0 & 0 & 0 \\ 0 + 3984.375 & -83.008 & 0 & 3984.375 \\ -1770.83 & 0 & -2390.625 & 0 \\ 170000 + 255000 & -3984.375 & 0 & 127500 \\ -3984.375 & 83.008 & 0 & -3984.375 \\ 0 & 0 & 2390.625 & 0 \\ 127500 & -3984.375 & 0 & 255000 \end{bmatrix}.$$

The load matrix is

$$\{F\}^{(1)} = \begin{bmatrix} 0 \\ -\frac{wL_1}{2} \\ -\frac{wL_1^2}{12} \\ 0 \\ -\frac{wL_1}{2} \\ \frac{wL_1^2}{12} \end{bmatrix} = \begin{bmatrix} 0 \\ -\frac{1000 \times 12}{2} \\ -\frac{1000 \times 12^2 \times 12}{12} \\ 0 \\ -\frac{1000 \times 12}{2} \\ \frac{1000 \times 12^2 \times 12}{12} \end{bmatrix} = \begin{bmatrix} 0 \\ -6000 \\ -144000 \\ 0 \\ -6000 \\ 144000 \end{bmatrix}.$$

Using boundary condition as $U_{11} = U_{12} = U_{13} = U_{31} = U_{32} = U_{33} = 0$, the global stiffness matrix is reduced to the following 3×3 matrix as given below:

$$10^3 \begin{bmatrix} 1676.758 & 0 & 3984.375 \\ 0 & 2415.22 & -1770.83 \\ 3984.375 & -1770.83 & 425000 \end{bmatrix} \begin{Bmatrix} U_{21} \\ U_{22} \\ U_{23} \end{Bmatrix} = \begin{bmatrix} 0 \\ -6000 \\ 144000 \end{bmatrix}.$$

Solving the above equations, we get

$$[U] = \begin{bmatrix} 0 \\ 0 \\ 0 \\ -0.00080082 \\ -0.00223715 \\ -0.00033701 \\ 0 \\ 0 \\ 0 \end{bmatrix} \text{ in.}$$

5.8 EXERCISE PROBLEMS

5.1. A triangular truss of equal sized 304 Stainless Steel bars is constructed with the bottom left node on a roller and the bottom right node fixed to a vertical wall (see Fig. 5.17). Each member has a cross sectional area of 5 cm² and a modulus of elasticity of 190 GPa. Determine the displacements at the nodes.

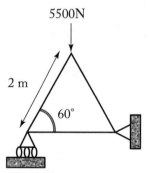

Figure 5.17: Problem 5.1.

5.2. A truss made from 304 Stainless Steel bars is constructed, as shown in Fig. 5.18. Each member has a cross sectional area of 4.2 cm² and a modulus of elasticity of 190 GPa. Determine the displacements at the node where the force is applied.

5.3. A bridge truss is made from aluminum ($E = 70$ GPa) and subjected to forces, as shown in Fig. 5.19. Each member has a cross sectional area of 20 cm². Determine the displacements on all nodes.

5.4. For the truss shown in Fig. 5.20, using two-element finite element model, determine the displacement of node 1 and the axial force in each element. A force of $P = 1000$ kN is

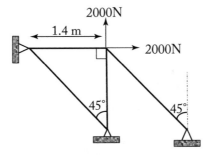

Figure 5.18: Problem 5.2.

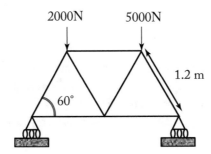

Figure 5.19: Problem 5.3.

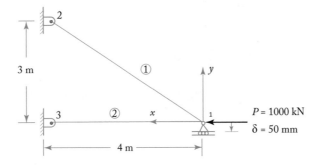

Figure 5.20: Problem 5.4.

applied at node 1 in the positive x direction while node 1 settles an amount $\delta = 50$ mm in the negative y direction. Assume $E = 210$ GPa and $A = 6.0 \times 10^{-4}$ m^2 for each element.

5.5. For the truss shown in Fig. 5.21, determine the deflections at point 3 and the axial stress in each member of the truss. Assume $EA = 30.0 \times 10^3$ Ksi and length, $l = 10$ in.

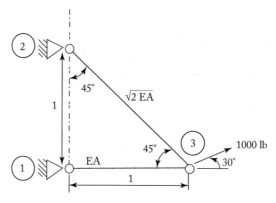

Figure 5.21: Problem 5.5.

5.6. Find the nodal displacements of the truss depicted in Fig. 5.22 with load $P = 1000$ N applied vertically downward at node 4. The elastic modulus for all members is $E = 2 \times 10^6$ N/cm^2. The geometric properties for elements #1 and #2 are ($A = 2.0$ cm^2; $L = 70.7$ cm); element #3 ($A = 1.0$ cm^2; $L = 158.1$ cm); and element #4 ($A = 1.0$ cm^2; $L = 141.4$ cm).

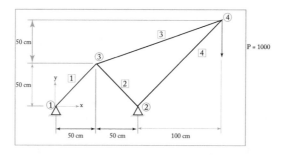

Figure 5.22: Problem 5.6.

5.7. For the beam shown in Fig. 5.23, find the maximum deflection and bending stress. Assume $E = 200$ GPa.

5.8. For the beam shown in Fig. 5.24, find the maximum deflection and bending stress. Assume $E = 200$ GPa.

5.9. Determine a rectangular structure beam to support the loads shown in Fig. 5.25. The bending stress should not exceed 190 MPa.

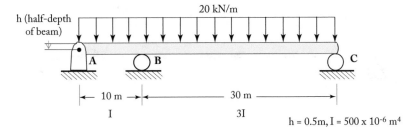

Figure 5.23: Problem 5.7.

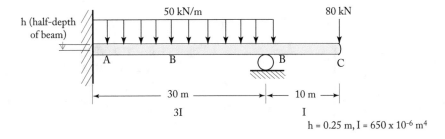

Figure 5.24: Problem 5.8.

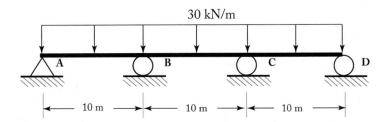

Figure 5.25: Problem 5.9.

5.10. Determine a rectangular structure beam to support the loads shown in Fig. 5.26. The bending stress should not exceed 190 MPa.

5.11. A frame structure made from steel ($E = 210$ GPa) is subjected to loads as shown in Fig. 5.27. Assume all the frame members have same area and moment of inertia. Find the unknown deflections and rotations as well as reaction forces.

5.12. A frame structure made from steel ($E = 210$ GPa) is subjected to loads, as shown in Fig. 5.28. Assume all the frame members have the same area and moment of inertia. Find the unknown deflections and rotations as well as reaction forces.

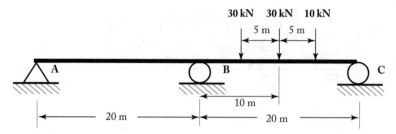

Figure 5.26: Problem 5.10.

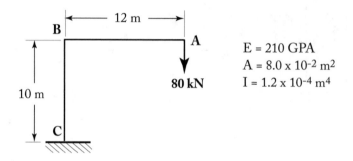

Figure 5.27: Problem 5.11.

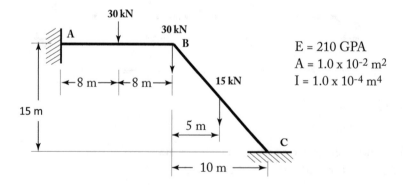

Figure 5.28: Problem 5.12.

5.13. A frame structure made from steel is subjected to loads as shown in Fig. 5.29. Assume all the frame members have same area, moment of inertia and polar moment of inertia. Find the unknown deflections and rotations as well as reaction forces.

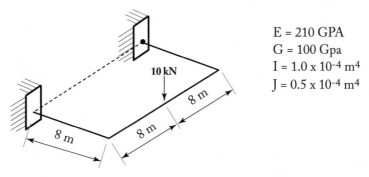

E = 210 GPA
G = 100 Gpa
I = 1.0 x 10⁻⁴ m⁴
J = 0.5 x 10⁻⁴ m⁴

Figure 5.29: Problem 5.13.

5.14. For the frame structure made from steel and subjected to loads as shown in Fig. 5.30, assume all the frame members have same area, moment of inertia and polar moment of inertia. Find the unknown deflections and rotations as well as reaction forces.

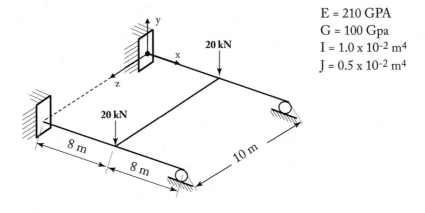

E = 210 GPA
G = 100 Gpa
I = 1.0 x 10⁻² m⁴
J = 0.5 x 10⁻² m⁴

Figure 5.30: Problem 5.14.

CHAPTER 6

1D FEA Using ANSYS

After reading this chapter, you will be able to:

- analyze 1D stress, heat transfer, and fluid problems with ANSYS;

- follow a step by step procedure for conducting FEA; and

- find a solution and display analysis results.

6.1 OVERVIEW

This chapter introduces the tools and techniques that are an integral part of the ANSYS software for analyzing 1D examples involving solid, heat transfer, fluid, and free vibration. The analysis procedure using ANSYS are explained in a table format and illustrated with screen shots as needed, to facilitate better understanding.

6.2 ELEMENTS IN ANSYS FOR 1D ANALYSIS

The examples presented in the next section are based on 1-D ANSYS elements with two or three nodes. These elements are summarized in Table 6.1.

6.3 AXIAL BAR UNDER TENSILE LOAD

Consider a simple example of a cylindrical rod of length l and cross-sectional area A, subjected to a tensile force F, as shown in Fig. 6.1. Find the deflection and stress in the bar. The dimensions are given as length, $l = 10$ in; diameter, $d = 1$ in; cross-sectional area, $A = 0.7854$ in^2 and the applied force, $F = 5,000$ lb. The material properties are, Young's Modulus, $E = 30 \times 10^6$ psi and Poisson's ratio $= 0.29$ and yield strength, $\sigma_y = 30$ ksi.

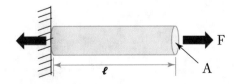

Figure 6.1: Axial bar under tensile load.

Table 6.1: Typical ANSYS elements for 1D analysis

ANSYS Element	Nodes with DOF	Degrees of Freedom (DOF)
1D Bar/Truss: LINK180 Structural, 2-Node		UX, UY, UZ
Beam: BEAM188 Structural, 2-Node		UX, UY, UZ, ROTX, ROTY, ROTZ
Beam: BEAM189 Structural, 3-Node		UX, UY, UZ, ROTX, ROTY, ROTZ
Conduction Heat Transfer: LINK33, 2 Node	T_i o———o T_j	TEMP
Convection Heat Transfer: LINK34, 2-Node	T_i o $\sim\!\!\sim\!\!\sim$ o T_j	TEMP
Frame: PIPE288, 2-Node		UX, UY, UZ, ROTX, ROTY, ROTZ

Solution: The ANSYS element LINK180 is used for the analysis to find deflection and stress in the bar. Table 6.2 shows a step-by-step procedure for analyzing the problem.

I. Pre-processing

1. *Specify element type.* Link element is the most appropriate as the bar is subjected to only axial loads (Fig. 6.2).

Table 6.2: ANSYS steps for analysis of 1D axial bar

I. Pre-Processing	Input
1. Specify element type:	
a. Preprocessor → Element Type → Add/Edit/Delete → Add..	Structural Mass, Link, 3D finite strain 180
2. Specify real constants:	
a. Preprocessor → Real Constants → Add/Edit/Delete → Add..	AREA <Cross-sectional area>
3. Define material properties:	
a. Preprocessor → Material Props → Material Models → Material Model Number 1	Linear, Elastic, Isotropic, EX <Young's modulus>, PRXY <Poisson's ratio>
4. Create model geometry	
a. Create nodes: Preprocessor → Modeling → Create → Nodes → In Active CS	NODE <Node number>, X,Y, Z <Location of node in active coordinate system>
b. Specify element attributes: Preprocessor → Modeling → Create → Elements → Elem Attributes	TYPE <1 LINK180>
c. Create elements: Preprocessor → Modeling → Create → Elements → Auto Numbered → Thru Nodes	Select nodes as endpoints of element
5. Apply boundary conditions	
a. Define displacements: Preprocessor → Loads → Define Loads → Apply → Structural → Displacement → On Nodes	Select fixed node
b. Define forces: Preprocessor → Loads → Define Loads → Apply → Structural → Force/Moment → On Nodes	Select node on which force is applied
II. SOLUTION: Solution → Solve → Current LS	
III. POST-PROCESSING	
1. Plot results	
a. Deformed shape: General Postprocessing → Plot Results → Deformed Shape	
2. List results	
a. Nodal solution: General Postprocessing → List Results → Nodal Solution	DOF Solution, Displacement vector sum
b. Element solution: General Postprocessing → List Results → Element Solution	Line Element Results, Element Results

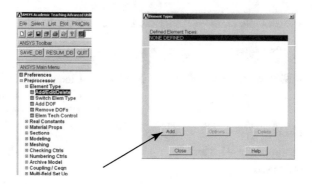

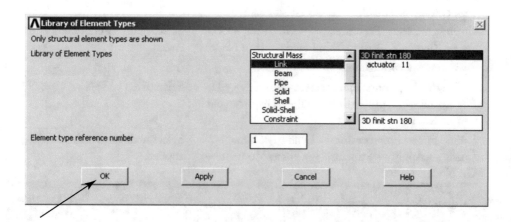

Figure 6.2: Specify element type.

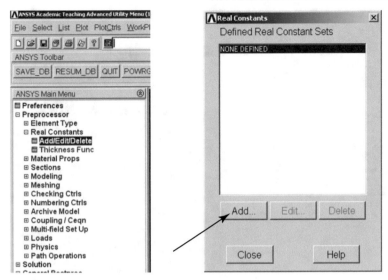

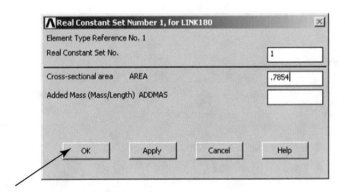

Figure 6.3: Specify real constants.

2. *Specify real constants.* In this case, only the cross-sectional area is needed (Fig. 6.3).

3. *Define material properties.* Assume a linear-elastic isotropic model; enter the Young's modulus (EX), and Poisson's ratio (PRXY) (Fig. 6.4).

4. (a) *Create model geometry: Nodes.* In this case, the nodes are the end-points of the bar (Fig. 6.5).

   The nodes should appear as in the screen shot (Fig. 6.6).

4. (b) *Create model geometry: Element attributes.* This will assign the material properties and real constants to the Link element (Fig. 6.7).

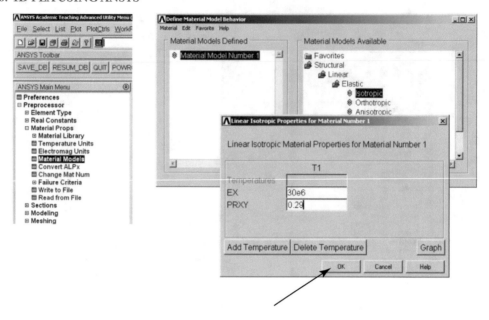

Figure 6.4: Define material properties.

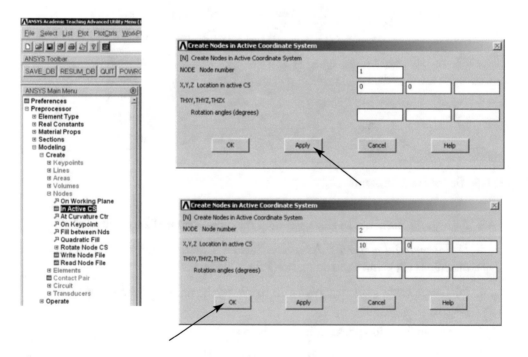

Figure 6.5: Create model geometry.

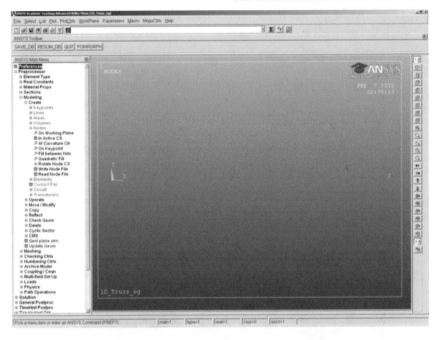

Figure 6.6: The nodes.

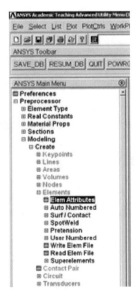

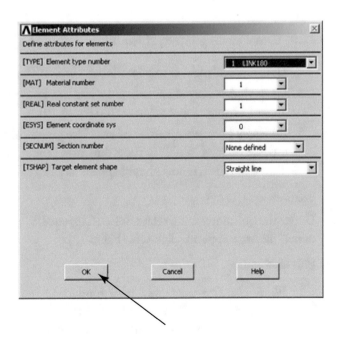

Figure 6.7: Create model geometry: Element attributes.

4. (c) *Create model geometry: Elements.* This will create an element with end-points through the nodes (Fig. 6.8).

 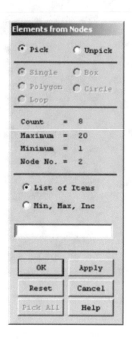

Figure 6.8: Create model geometry: Elements.

Use graphical picking to select nodes 1 and 2. The element should appear as in Fig. 6.9.

5. (a) *Apply boundary conditions: Displacement.* Use graphical picking to select Node 1 and apply a Displacement value of "0" (Fig. 6.10).

5. (b) *Apply boundary conditions: Force.* Use graphical picking to select Node 2 and apply a Force value of "5000" in the x direction (Fig. 6.11).

The constrained model should appear as the screen shot in Fig. 6.12.

II. Solution

Under the Solution menu, select Solve (Current LS, and click OK. When the "Solution is done!" message appears, click OK (Fig. 6.13).

III. Post-processing

1. *Plot results: Deformed shape.* It's a good idea to check the deformed shape of the model to see if the shape changes as is expected with the applied load (Fig. 6.14).

The bar should show elongation in the positive x direction, as in Fig. 6.15. Other useful plot types, such as contour and vector plots, will be discussed in subsequent examples.

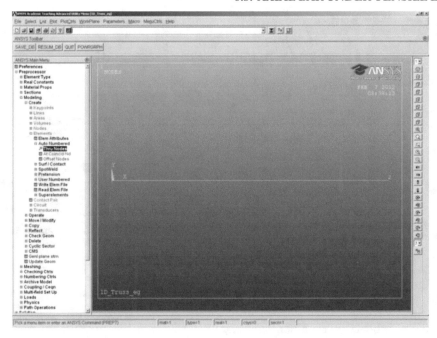

Figure 6.9: Nodes 1 and 2.

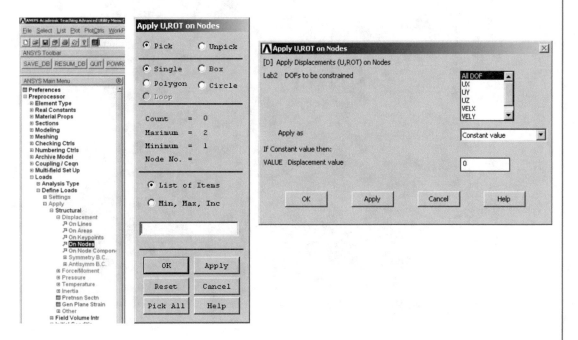

Figure 6.10: Apply boundary conditions: Displacement.

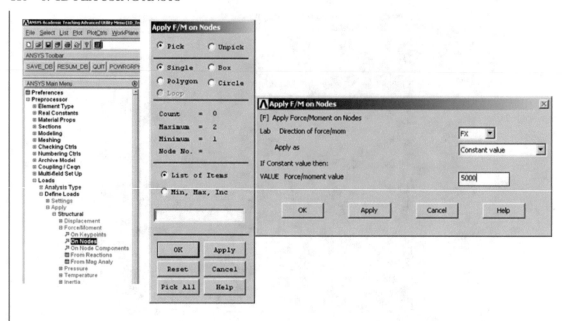

Figure 6.11: Apply boundary conditions: Force.

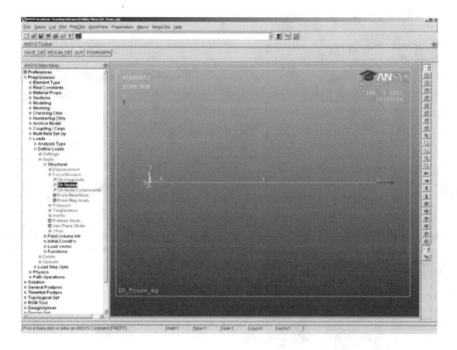

Figure 6.12: Constrained model.

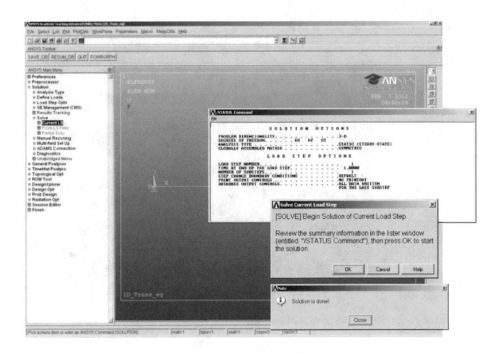

Figure 6.13: Solution menu.

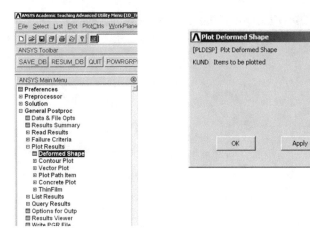

Figure 6.14: Plot results: Deformed shape.

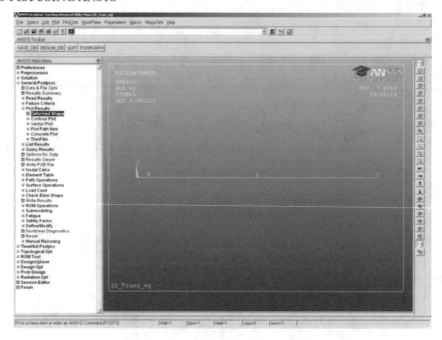

Figure 6.15: Elongation in the positive x direction.

2. (a) *List results: Nodal solution.* This will list the displacements of the individual nodes (Fig. 6.16).

2. (b) *List results: Element solution.* This gives the total stress in the element (Fig. 6.17).

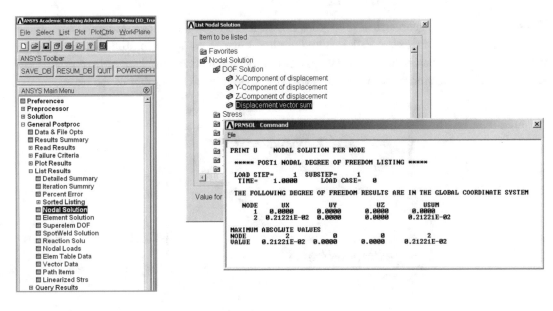

Figure 6.16: List results: Nodal solution.

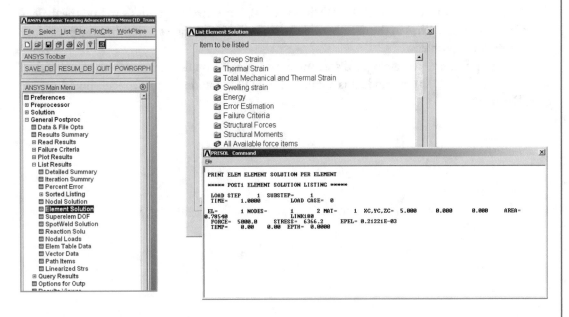

Figure 6.17: List results: Element solution.

6.4 TRUSS ANALYSIS–EXAMPLE #1

Consider a truss with geometrical properties shown in Fig. 6.18. The cross-sectional area, same for all the members of truss, $A = 1$ in^2; and applied force, $F = 1,000$ lb. The material properties are: Young's Modulus, $E = 30 \times 10^6$ psi; Poisson's ratio $= 0.29$; yield strength, $\sigma_y = 30$ ksi. Determine the deflection of each joint and also the stresses in each member of the truss.

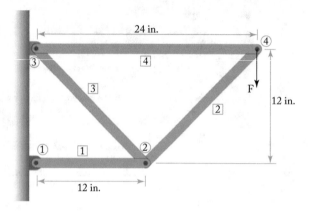

Figure 6.18: Truss example #1.

Solution: The ANSYS element LINK 3D finit stn 180, is used for the analysis to find the deflection of each joint and stress in the members of the truss. Table 6.3 shows a step-by-step procedure for analyzing the problem.

For step 5 in pre-processing, a displacement value of "0" on nodes 1 and 3 (the fixed nodes) and a force FY of "-1000" on node 3 is applied. The fully constrained truss model should appear as in Fig. 6.19.

Table 6.3: ANSYS solution steps for plane truss analysis *(Continues.)*

I. Pre-Processing	Input
1. Specify element type:	
a. Preprocessor → Element Type → Add/Edit/Delete → Add..	Structural Mass, Link, 3D finite strain 180
2. Specify real constants:	
a. Preprocessor → Real Constants → Add/Edit/Delete → Add..	AREA <Cross-sectional area>
3. Define material properties:	
a. Preprocessor → Material Props → Material Models → Material Model Number 1	Linear, Elastic, Isotropic, EX <Young's modulus>, PRXY <Poisson's ratio>
4. Create model geometry	
a. Create nodes: Preprocessor → Modeling → Create → Nodes → In Active CS	NODE <Node number>, X,Y, Z <Location of node in active coordinate system>
b. Specify elements: Preprocessor → Modeling → Create → Elements → User Numbered → Thru Nodes	Select nodes as endpoints of element
5. Apply boundary conditions	
a. Define displacements: Preprocessor → Loads → Define Loads → Apply → Structural → Displacement → On Nodes	Select nodes with displacement constraints
b. Define forces: Preprocessor → Loads → Define Loads → Apply → Structural → Force/Moment → On Nodes	Select node on which forces are applied

Table 6.3: *(Continued.)* ANSYS solution steps for plane truss analysis

II. SOLUTION: Solution → Solve → Current LS	
III. POST-PROCESSING	
1. Plot results	
a. Deformed shape: General Postprocessing → Plot Results → Deformed Shape	
2. List results	
b. Nodal solution: General Postprocessing → List Results → Nodal Solution	DOF Solution, Displacement vector sum
c. Reaction solution: General Postprocessing → List Results → Reaction Solution	All items
3. Element tables	
a. Define force table: General Postprocessing → Element Table → Define Table →Add	Lab <enter user label, e.g., "axforce">, Item Comp Results data item <By sequence num>, <SMISC,1>
b. Define stress table: General Postprocessing → Element Table → Define Table → Add	Lab <enter user label, e.g., "axstress">, Item Comp Results data item <By sequence num>, <LS,1>
c. List element table: General Postprocessing → Element Table → List Elem Table	<select tables defined in previous step>

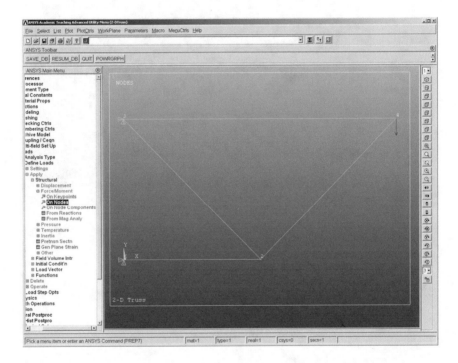

Figure 6.19: The fully constrained truss model.

The deformed shape of the truss and nodal displacement results are shown in Fig. 6.20.

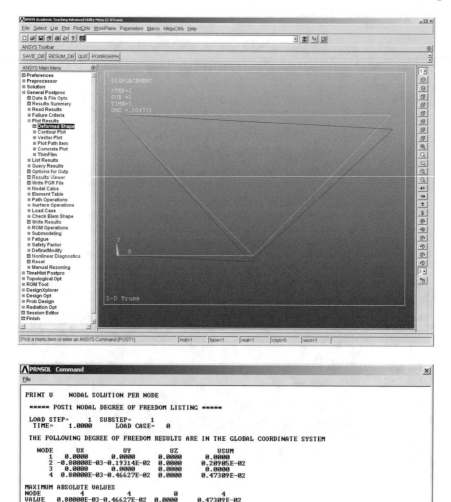

Figure 6.20: Deformed shape of the truss and nodal displacement results.

6.5   TRUSS ANALYSIS–EXAMPLE #2

Consider a water tower with a truss-type support structure shown in Fig. 6.21 for deflection analysis. It is assumed that the only applied loads are from the weight of the tank at the top of the structure. The cross-sectional area, $A = 177$ cm^2; is same for all the members of the truss and a force (each upper node), $F = 5{,}000$ N is applied. Use Young's Modulus, $E = 200$ GPa; and Poisson's ratio $= 0.27$ for the analysis.

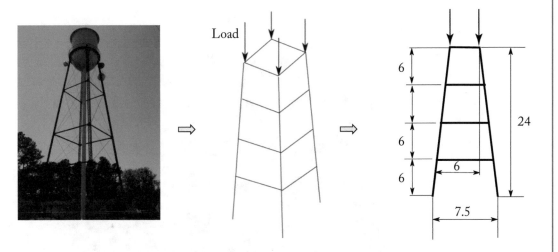

Figure 6.21: Water tower model (dimensions in meters).

Solution: Due to symmetry of the problem, we can consider only one of the four sides of the structure for analysis. Following the steps in Table 6.3, specify the element type, real constants, and material modes, and then create the nodes for the model according to Table 6.4, and connect the nodes with elements as in the previous example.

Table 6.4: Water tank model nodal coordinates

Node no.	X (m)	Y (m)
1	0	0
2	7.5	0
3	1.5	24
4	6	24
5	0.375	6
6	0.75	12
7	1	18
8	7.125	6
9	6.75	12
10	6.5	18

After creating the nodes and elements, apply the loads to the top two nodes and the zero displacement conditions to the bottom two nodes. The model should appear as the screen shot in Fig. 6.22.

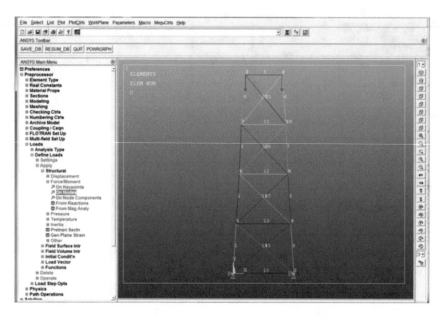

Figure 6.22: Screen shot of the model.

After running the Solution, the Contour Plot (Displacement vector sum), with the deformed shape, is shown in the screen shot as in Fig. 6.23.

6.6   SPACE TRUSS ANALYSIS–EXAMPLE

Consider a space truss shown in Fig. 6.24, with 12 elements and 6 nodes, connected to a vertical wall by a ball-and-socket joint at node 4, and pivoting sliding joints at nodes 2 and 3. Forces are applied at nodes 5 and 6 in the positive x and negative y directions, respectively. The cross-sectional area, $A = 1.5$ in^2; is same for all the members of the truss. Use Young's Modulus, $E = 11 \times 10^6$ psi; and Poisson's ratio $= 0.30$ for the analysis. Determine the deflection of the truss and the reactions.

Note that the ball-and-socket support at node 1 will prevent translation of the joint (allowing only rotation), while the sliders allow for limited translation in the y- and z-directions at node 2, and in the z-direction at node 3. This will be important when applying the displacement constraints for those nodes.

Solution: The steps for for completing the analysis using ANSYS for the space truss are the same as for the 2D truss examples (see Tables 6.2 and 6.3).

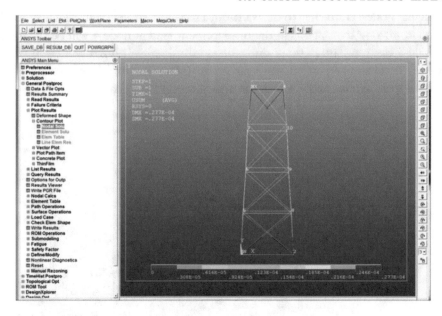

Figure 6.23: Screen after running the solution.

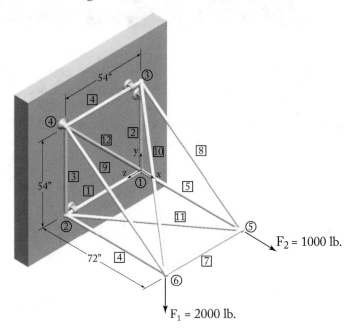

Figure 6.24: Space truss assembly affixed to a vertical wall. Circled numbers are node numbers and boxed numbers are element numbers.

Recall that node 4 is attached to the wall with a ball-and-socket joint, allowing only rotation about the node with no translation, so the displacement load at node 4 will be "0" for UX, UY, and UZ. Node 3 can translate in the z direction, so displacement loads will be "0" for UX and UY, and node 2 is constrained only in the x direction ($UX = 0$). The fully constrained space truss model should appear similar to the screen shots, as given in Fig. 6.25.

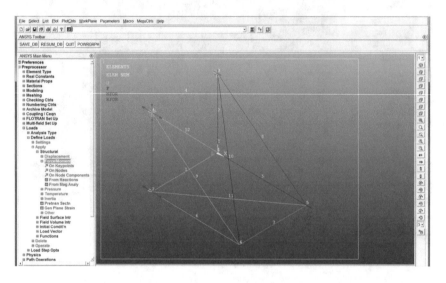

Figure 6.25: Fully constrained space truss model.

The deformed shape of the truss and nodal displacement and reaction results are shown in Fig. 6.26.

Element tables can be generated in order to obtain axial forces and stresses for each element, using the menu paths. The resulting table is shown in Fig. 6.27.

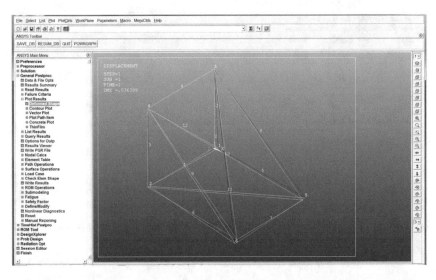

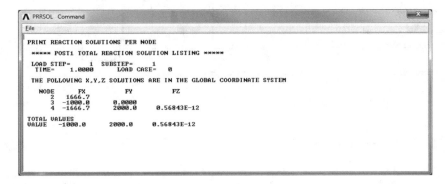

Figure 6.26: Deformed shape of the truss and nodal displacement and reaction results.

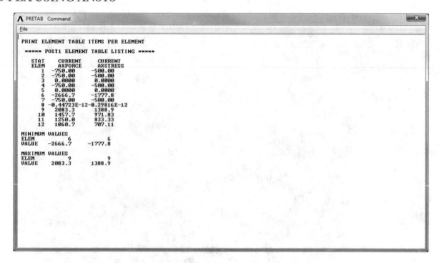

Figure 6.27: Element tables can be generated.

6.7   BEAM ANALYSIS–EXAMPLE

Consider a beam shown in Fig. 6.28, which depicts a W16x31 I-Beam with both concentrated and distributed loads applied. The beam is made from steel, with a Young's modulus of 30×10^6 psi and a Poisson's ratio of 0.29. The objective is to find the displacement of the I-beam under the applied load using ANSYS software.

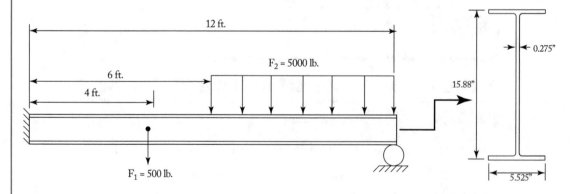

Figure 6.28: Beam example for analysis.

Solution: The ANSYS Beam 2-Node 188 element is used for the analysis. Table 6.5 shows the ANSYS steps for analyzing the beam problem.

Table 6.5: ANSYS steps for beam structure analysis *(Continues.)*

I. Pre-Processing	Input
1. Specify element type:	
a. Preprocessor → Element Type → Add/Edit/Delete → Add	Structural Mass, 3D finite strain, 2 node 188
2. Specify beam sections:	
a. Preprocessor → Sections → Beam → Common Sections	Enter beam section properties using the beam tool
3. Define material properties:	
a. Preprocessor → Material Props → Material Models → Material Model Number 1	Linear, Elastic, Isotropic, EX <Young's modulus>, PRXY <Poisson's ratio>
4. Create model geometry	
a. Create nodes: Preprocessor → Modeling → Create → Nodes → In Active CS	NODE <Node number>, X,Y, Z <Location of node in active coordinate system>
b. Create elements: Preprocessor → Modeling → Create → Elements → User Numbered → Thru Nodes	Select nodes as endpoints of element
5. Apply boundary conditions	
a. Define displacements: Preprocessor → Loads → Define Loads → Apply → Structural → Displacement → On Nodes	Select nodes with displacement constraints
b. Define forces: Preprocessor → Loads → Define Loads → Apply → Structural → Force/Moment → On Nodes	Select node on which forces are applied
c. Define pressure: Preprocessor → Loads → Define Loads → Apply → Structural → Pressure → On Elements	Select element on which distributed loads are applied

Table 6.5: *(Continued.)* ANSYS steps for beam structure analysis

II. SOLUTION: Solution → Solve → Current LS	
III. POST-PROCESSING	
1. Plot results	
a. Deformed shape: General Postprocessing → Plot Results → Deformed Shape	
2. List results	
b. Nodal solution: General Postprocessing → List Results → Nodal Solution	DOF Solution, Displacement vector sum
c. Reaction solution: General Postprocessing → List Results → Reaction Solution	All items
3. Element tables	
a. Define force table: General Postprocessing → Element Table → Define Table →Add	Lab <enter user label, e.g., "axforce">, Item Comp Results data item <By sequence num>, <SMISC,1>
b. Define stress table: General Postprocessing → Element Table → Define Table → Add	Lab <enter user label, e.g., "axstress">, Item Comp Results data item <By sequence num>, <LS,1>
c. List element table: General Postprocessing → Element Table → List Element Table	<select tables defined in previous step>

The beam section is specified using the Beam Tool (Preprocessor → Sections → Beam → Common Sections). To build the model, create nodes at the endpoints of the beam and at the loading points, and then create elements connecting the nodes.

For boundary conditions, apply zero displacement for all DOF at the node at the left end (fixed) of the beam, and zero displacement in the y direction at the right end (rolling support). The 500 lb point force is applied to the node at $x = 48$ in, and a pressure of 12.57 lb/in^2 (5,000 lb/area of applied load) is applied to the rightmost element. The fully constrained model should appear similar to the screen shot in Fig. 6.29.

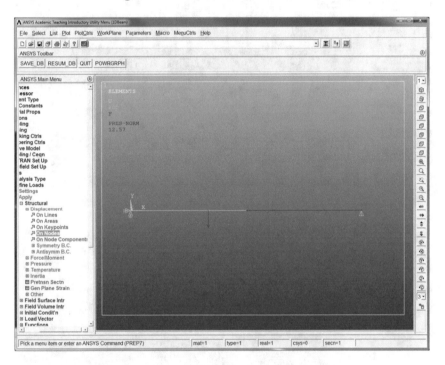

Figure 6.29: Fully constrained model.

The results for deformation and rotation are shown in Fig. 6.30.

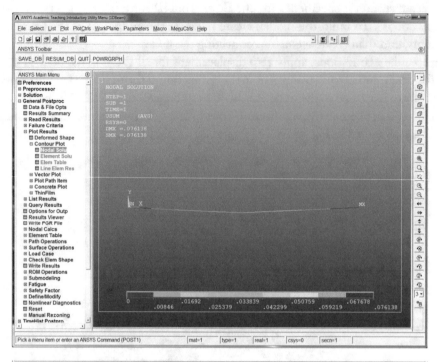

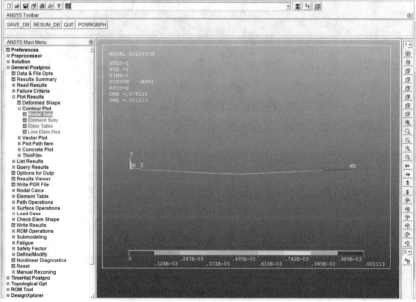

Figure 6.30: Results for deformation and rotation.

6.8 FRAME ANALYSIS–EXAMPLE

Consider a gantry crane for finite element analysis: where the frame is a 2D or 3D assembly of beam elements along with the dimensions shown below. Figure 6.31 depicts a gantry crane. The upper horizontal member is an I-beam, supported by two vertical rectangular tubes. The hoist is mounted on wheels, enabling it to roll along the length of the I-beam.

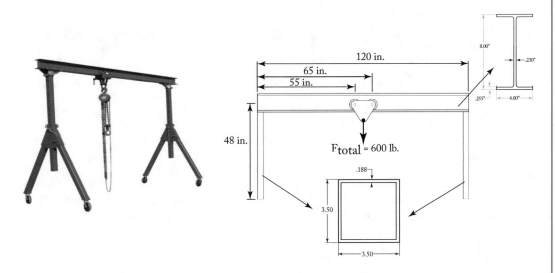

Figure 6.31: A gantry crane with chain hoist (www.northerntool.com) modeled as 2D frame for analysis.

For simplicity, only the upper part of the crane (from the attachment point of the vertical supports upwards) will be modeled. The objective is to find the displacement of the I-beam under a load applied in the middle of the beam. The vertical members are steel tubes with outer dimensions of 3 1/2 × 3 1/2 in and a wall thickness of 0.188 in. The horizontal I-Beam is W8×13. All members are steel, with a Young's modulus of 30×10^6 psi and Poisson's ratio of 0.29. Since the 600 lbs load is carried by the wheels, which are spaced 10 in apart, the applied load should be 300 lbs at each of the wheels (at $x = 55$ in and $x = 65$ in).

Solution: The BEAM188 2-node beam element is used in the analysis. The steps for for completing the analysis for the frame using ANSYS software are summarized in Table 6.6.

To define the sections, use the Beam tool (Preprocessor → Sections → Beam → Common Sections). Enter the data for each section type, as shown in Fig. 6.32.

To build the model, create keypoints at the ends of the beams ({0, 0}, {0, 48}, {120, 48}, and {120, 0}), at the load application points ({55, 48}) and ({65, 48}). Connect the keypoints with lines, and apply the mesh to the lines, applying Section 1 to the three lines representing the I-Beam in Mesh Attributes, and Section 2 to the two vertical lines. DOF constraints of "0"

Table 6.6: ANSYS steps for frame analysis *(Continues.)*

I. Pre-Processing	Input
1. Specify element type:	
a. Preprocessor → Element Type → Add/Edit/Delete → Add	Structural Mass, Beam, 3D finite strain, 2 node 188
2. Specify beam sections:	
a. Preprocessor → Sections → Beam → Common Sections	Enter beam section properties using the beam tool
3. Define material properties:	
a. Preprocessor → Material Props → Material Models → Material Model Number 1	Linear, Elastic, Isotropic, EX <Young's modulus>, PRXY <Poisson's ratio>
4. Create model geometry	
a. Create nodes: Preprocessor → Modeling → Create → Nodes → In Active CS	NODE <Node number>, X,Y, Z <Location of node in active coordinate system>
b. Create elements: Preprocessor → Modeling → Create → Elements → User Numbered → Thru Nodes	Select nodes as endpoints of element
5. Apply boundary conditions	
a. Define displacements: Preprocessor → Loads → Define Loads → Apply → Structural → Displacement → On Nodes	Select nodes with displacement constraints
b. Define forces: Preprocessor → Loads → Define Loads → Apply → Structural → Force/Moment → On Nodes	Select nodes on which forces are applied

Table 6.6: *(Continued.)* ANSYS steps for frame analysis

II. SOLUTION: Solution → Solve → Current LS	
III. POST-PROCESSING	
1. Plot results	
a. Deformed shape: General Postprocessing → Plot Results → Deformed Shape	
2. List results	
a. Nodal solution: General Postprocessing → List Results → Nodal Solution	DOF Solution, Displacement vector sum
b. Reaction solution: General Postprocessing → List Results → Reaction Solution	All items
3. Element tables	
a. Define force table: General Postprocessing → Element Table → Define Table →Add	Lab <enter user label, e.g., "axforce">, Item Comp Results data item <By sequence num>, <SMISC,1>
b. Define stress table: General Postprocessing → Element Table → Define Table → Add	Lab <enter user label, e.g., "axstress">, Item Comp Results data item <By sequence num>, <LS,1>
c. List element table: General Postprocessing → Element Table → List Element Table	<select tables defined in previous step>

should be applied to the bottom two keypoints, and a Force of $FY = -300$ to the loading points. The constrained model should appear similar to the screen shot in Fig. 6.33.

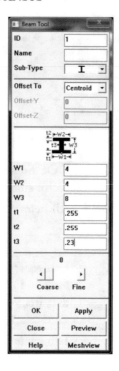
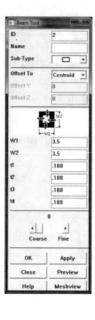

Figure 6.32: Enter the data for each section type.

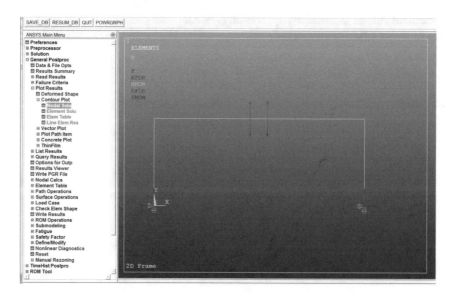

Figure 6.33: Constrained model.

After running the solution, the displacement contour plot should appear, as in Fig. 6.34.

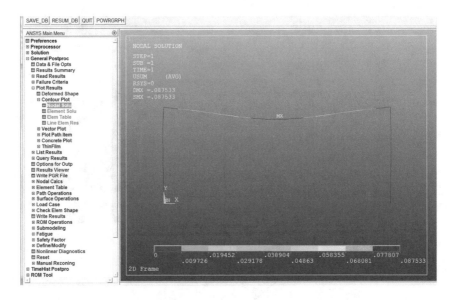

Figure 6.34: Displacement contour plot.

6.9   STATIC AND MODAL ANALYSIS OF A 3D FRAME

6.9.1   STRUCTURAL ANALYSIS

This example extends the frame structural analysis to three dimensions. Figure 6.35 shows a frame consisting of round tubular sections of various sizes supporting traffic signals.

It will be assumed that the only forces acting on the frame are the weights of the signals, represented by the forces P1–P4. The material properties for ASTM A36 are: Young's Modulus, $E = 200$ GPa; Poisson's ratio $= 0.26$; and density $= 7850$ kg/m^3. There are three different sizes for the sections, which are given in Table 6.7.

Solution: The BEAM188 2-node beam element will again be used. The inner and outer radii of each of the three sections are calculated using the data from Table 6.7. Also be sure to enter the density of the beam material, as it will be needed for the modal analysis in the next section. The steps for for completing the frame analysis with ANSYS software are summarized in Table 6.8.

To create the model, first use the data in Table 6.9 to generate the nodes, then create elements connecting the nodes. Use the Element Attributes tool to assign the correct section number to each of the elements prior to creating the element.

After creating the elements, the model should appear similar as in Fig. 6.36.

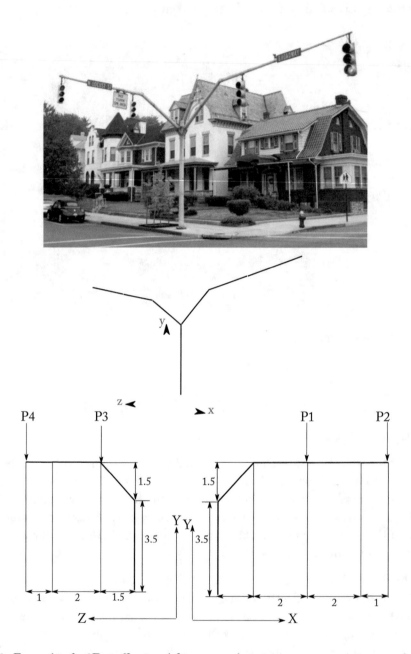

Figure 6.35: Example of a 3D traffic signal frame; numbers in squares are section numbers; dimensions are in meters.

Table 6.7: Frame section sizes for analysis

Section No.	Outside Diameter (m)	Wall thickness (m)
1	0.40	0.07
2	0.25	0.05
3	0.25	0.05
4	0.20	0.04
5	0.20	0.04

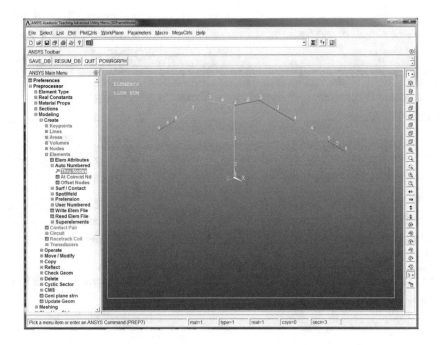

Figure 6.36: Model, after creating the elements.

Table 6.8: Steps for analyzing 3D frame using ANSYS *(Continues.)*

I. Pre-Processing	Input
1. Specify element type:	
a. Preprocessor → Element Type → Add/Edit/Delete → Add	Structural Mass, Beam, 3D finite strain, 2 node 188
2. Specify beam sections:	
a. Preprocessor → Sections → Beam → Common Sections	Enter beam section properties using the beam tool
3. Define material properties:	
a. Preprocessor → Material Props → Material Models → Material Model Number 1	Structrual, Linear, Elastic, Isotropic, EX <Young's modulus>, PRXY <Poisson's ratio>; Structural, Density, DENS <Material density>
4. Create model geometry	
a. Create nodes: Preprocessor → Modeling → Create → Nodes → In Active CS	NODE <Node number>, X,Y, Z <Location of node in active coordinate system>
b. Create elements: Preprocessor → Modeling → Create → Elements → User Numbered → Thru Nodes	Select nodes as endpoints of element
5. Apply boundary conditions	
a. Define displacements: Preprocessor → Loads → Define Loads → Apply → Structural → Displacement → On Nodes	Select nodes with displacement constraints
b. Define forces: Preprocessor → Loads → Define Loads → Apply → Structural → Force/Moment → On Nodes	Select nodes on which forces are applied

Table 6.8: *(Continued.)* Steps for analyzing 3D frame using ANSYS

II. SOLUTION: Solution → Solve → Current LS III. POST-PROCESSING	
1. Plot results	
a. Deformed shape: General Postprocessing → Plot Results → Deformed Shape	
2. List results	
a. Nodal solution: General Postprocessing → List Results → Nodal Solution	DOF Solution, Displacement vector sum
b. Reaction solution: General Postprocessing → List Results → Reaction Solution	All items
3. Element tables	
a. Define force table: General Postprocessing → Element Table → Define Table →Add	Lab <enter user label, e.g., "axforce">, Item Comp Results data item <By sequence num>, <SMISC,1>
b. Define stress table: General Postprocessing → Element Table → Define Table → Add	Lab <enter user label, e.g., "axstress">, Item Comp Results data item <By sequence num>, <LS,1>
c. List element table: General Postprocessing → Element Table → List Element Table	<select tables defined in previous step>

Table 6.9: Coordinates of frame nodes for creating the model

Node no.	X	Y	Z
1	0	0	0
2	0	3.5	0
3	1.5	4.5	0
4	3.5	4.5	0
5	5.5	4.5	0
6	6.5	4.5	0
7	0	4.5	1.5
8	0	4.5	3.5
9	0	4.5	4.5

For boundary conditions, apply a constant value of zero displacement for all DOF to node "1" (at the origin), and Forces of -75 N in the y direction to the signal mounting points (nodes 4, 6, 7, and 9). The fully constrained model should appear similar to the screen shot shown in Fig. 6.37.

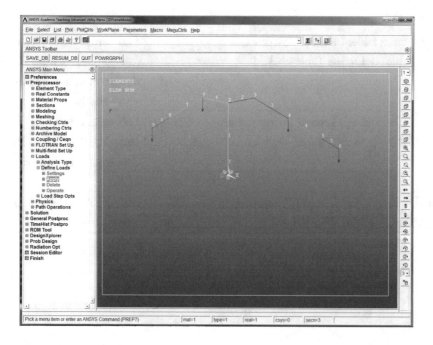

Figure 6.37: Fully constrained model.

The results of displacement vector sum and the von-Mises stresses from the analysis are shown in Fig. 6.38.

6.9.2   MODAL ANALYSIS

For the modal (vibration) analysis, the frame model from the previous structural example will be used. To start the modal analysis, use the menu path: Solution → Analysis Type → New Analysis → Modal. Next, set the analysis options using the path: Solution → Analysis Type → Analysis Options, and enter the options as shown in Fig. 6.39.

Initiate the solution (Solution → Solve → Current LS), then select Results Summary from the General Postproc menu, which will display the first five natural frequencies. The results should be similar to the screen shots, as shown in Fig. 6.40.

The first mode shape is displayed by selecting: General Postproc → Read Results → First Set, then General Postproc → Plot Results → Deformed Shape → Def + undeformed. The first mode shape is shown in Fig. 6.41.

Subsequent mode shapes can be displayed by selecting: General Postproc → Read Results → Next Set, then replotting the deformed shape. The third and fifth mode shapes are shown in Figs. 6.42 and 6.43.

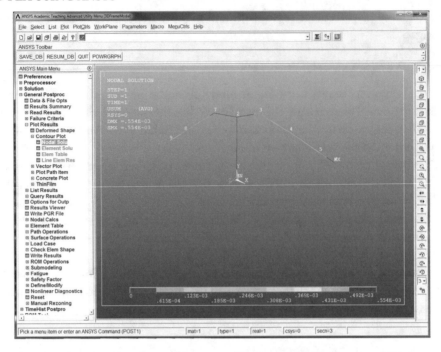

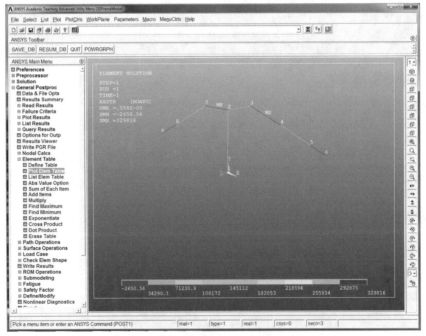

Figure 6.38: Results.

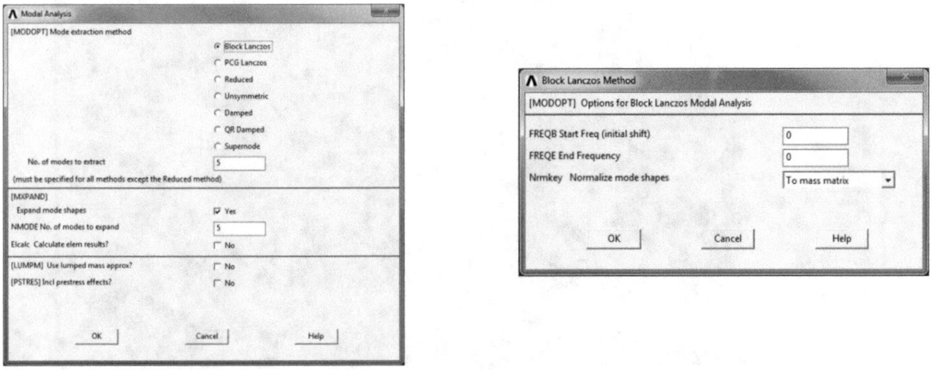

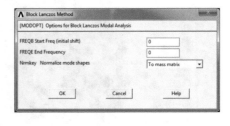

Figure 6.39: Modal analysis.

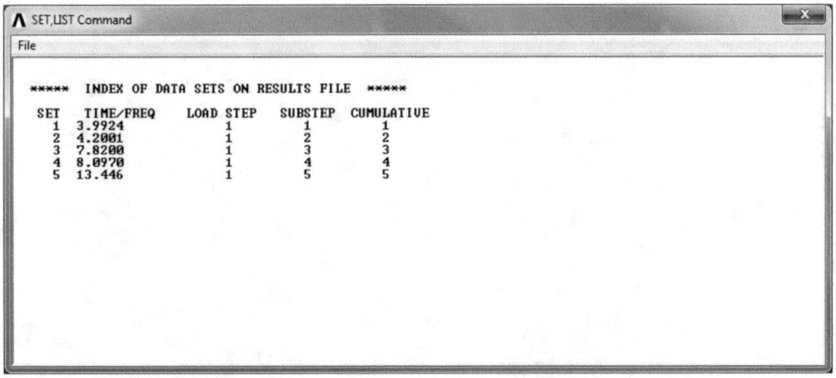

Figure 6.40: Results summary.

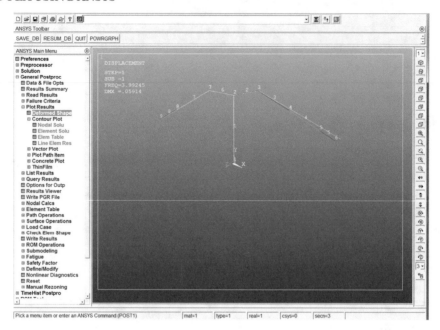

Figure 6.41: First mode shape.

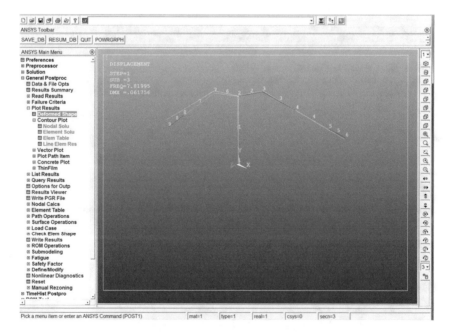

Figure 6.42: Third mode shape.

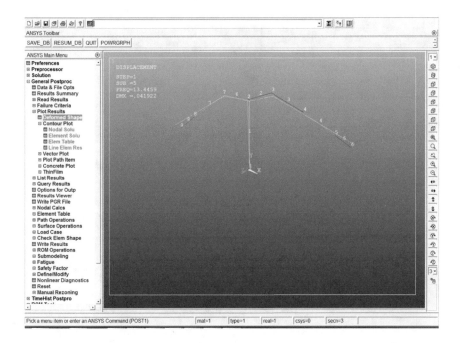

Figure 6.43: Fifth mode shape.

6.10   1D HEAT TRANSFER ANALYSIS–EXAMPLE

This example will illustrate the analysis of 1D thermal conduction through a vertical wall, with convection acting on the vertical surfaces. The wall shown in Fig. 6.44 separates two regions having air temperatures of $T_{\infty,1}$ and $T_{\infty,2}$, which are known. Also, given are the thermal conductivity, k, convective coefficients, h_1 and h_2, length (thickness of wall), L, and surface area of each side of the wall, A. The system is assumed to be in steady-state. ANSYS will be used to find the temperature at each surface, $T_{S,1}$ and $T_{S,2}$, and the interior temperature T_i. The properties are: thermal conductivity, $k = 52$ W/mK; convective coefficient, $h = 10$ W/m^2 K; and wall thickness, $L = 0.5$ m; $T_{\infty,1} = 150°C$; $T_{\infty,2} = 25°C$ and wall surface area (each side), $A = 1$ m^2.

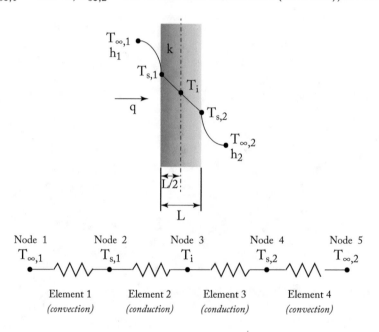

Figure 6.44: Vertical wall, element, and node configuration.

Solution: For thermal analysis, appropriate ANSYS elements are LINK33 (conduction) and LINK34 (convection). The steps for for completing the analysis in ANSYS for the 1D Heat Transfer example are summarized in Table 6.10.

The screen shots of model nodes, finite element model and the table of nodal temperatures are in Fig. 6.45.

Table 6.10: ANSYS steps for 1D Heat transfer analysis

I. Pre-Processing	Input
1. Specify element type	
a. Conduction: Preprocessor → Element Type → Add/ Edit/Delete → Add	Thermal Mass, Solid, 8node 77
b. Convection: Preprocessor → Element Type → Add/ Edit/Delete → Add	Thermal Mass, Link, 3D conduction 33
2. Specify real constants:	
a. Area: Preprocessor → Real Constants → Add/Edit/ Delete → Add	Cross-sectional area <Area>
3. Define material properties:	
a. Preprocessor → Material Props → Material Models → Material Model Number 1	Thermal, Isotropic, KXX, <Thermal conductivity>
4. Create model geometry	
a. Create area: Preprocessor → Modeling → Create → Areas → Rectangle → By 2 Corners	WP X, WP Y <Location of origin>, Width, Height
b. Create mesh: Preprocessor → Meshing → Mesh → Areas→ Mapped → 3 or 4 sided	Pick All
c. Refine mesh: Preprocessor → Meshing → Modify Mesh → Refine At→ Areas	LEVEL <Level of refinement>
5. Apply boundary conditions	
a. Define convection: Preprocessor → Loads → Define Loads → Apply → Thermal → Convection → On Lines	Select lines representing surfaces, VALI<Convective coefficient>, VAL21 <T∞>
II. SOLUTION: Solution → Solve → Current LS	
III. POST-PROCESSING	
1. Plot results	
a. Contour Plot: General Postprocessing → Plot Re-sults → Contour Plot Nodal Solu	DOF Solution, Nodal Temperature
2. List results	
a. Nodal solution: General Postprocessing → List Re-sults → Nodal Solution	DOF Solution, Nodal Temperature

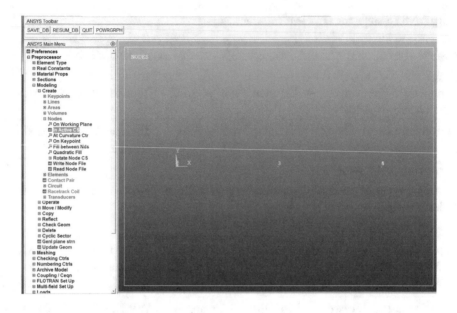

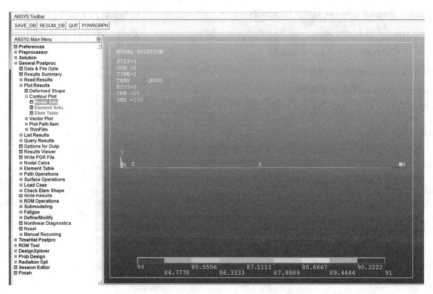

Figure 6.45: Screen shots of model nodes, finite element model, and the table of nodal temperatures *(Continues.)*

```
A PRNSOL Command                                                    x

File

PRINT TEMP NODAL SOLUTION PER NODE

 ***** POST1 NODAL DEGREE OF FREEDOM LISTING *****

  LOAD STEP=     1  SUBSTEP=     1
   TIME=     1.0000    LOAD CASE=   0

     NODE      TEMP
       1     150.00
       2      90.367
       3      87.500
       4      84.633
       5      25.000

 MAXIMUM ABSOLUTE VALUES
 NODE        1
 VALUE    150.00
```

Figure 6.45: *(Continued.)* Screen shots of model nodes, finite element model, and the table of nodal temperatures.

6.11   EXERCISE PROBLEMS

Use problems given in Chapters 4 and 5, solve them using ANSYS software and compare your results with those obtained by hand.

CHAPTER 7

2D Finite Element Analysis

After reading this chapter, you will be able to:

- understand 2D problems in engineering;

- define and find shape functions for 2D elements;

- define isoparametric formulation and higher order elements;

- analyze 2D solid, heat transfer, and fluid flow problems; and

- know about 2D elements in ANSYS and use them for analysis.

7.1 OVERVIEW

In previous chapters, we learned about 1D analysis of engineering problems. In general, all engineering applications require making approximations. Most practical components are 3D and we need to make assumptions to find a solution, especially for 2D for solving different problems. Specifically, an engineering problem assumed as 2D that can be solved using 2D finite elements is described in this chapter.

7.2 ELEMENT TYPES AND SHAPE FUNCTIONS

In general, a plane solid (2D geometry) can be modeled using a variety of 2D elements, as shown in Fig. 7.1. The elements may have different shapes (triangular, rectangular or quadrilateral) and different nodes. At each node, there will a DOF which depends on the type of problem we are interested in solving. For example, in heat transfer analysis, the nodal variable will be temperature at each node of the element. However, in solid mechanics problems, each node in an element may have two DOF (displacement in x and y directions). The first step in deriving the FE matrix equations is to assume interpolation functions describing the behavior of the element in terms of its nodal values. In order to illustrate 2D elements and their shape functions, typical triangular and rectangular elements are discussed in the next section.

7.2.1 TRIANGULAR ELEMENT

A triangular element is defined by three nodes at the three vertices of a triangle, as shown in Fig. 7.2. Consider, at each node, the dependent variable is temperature. The temperature distri-

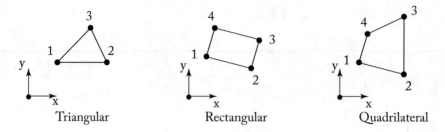

Figure 7.1: Types of 2D plane elements with different shape and number of nodes.

bution function over the triangular element, $T^{(e)}$ is represented as

$$T^{(e)} = b_1 + b_2 x + b_3 y,$$

where b_1, b_2, and b_3 are constants. In FEA, the constants are replaced with nodal values.

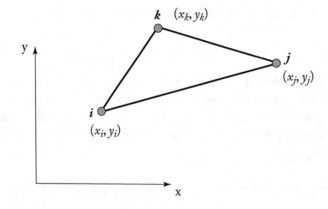

Figure 7.2: Three-node triangular element.

Using the following conditions:

$$@x = x_i \text{ and } y = y_i, \; T = T_i$$
$$@x = x_j \text{ and } y = y_j, \; T = T_j$$
$$@x = x_k \text{ and } y = y_k, \; T = T_k$$

the constants b_1, b_2, and b_3 can be determined. After finding the constants and substituting back into element temperature distribution function, we can simplify as

$$T^{(e)} = \begin{bmatrix} S_i & S_j & S_k \end{bmatrix} \left\{ \begin{array}{c} T_i \\ T_j \\ T_k \end{array} \right\}.$$

The S_i, S_j, and S_k are the shape functions, which are given as

$$S_i = \frac{1}{2A} \left(\alpha_i + \beta_i x + \delta_i y \right)$$

$$S_j = \frac{1}{2A} \left(\alpha_j + \beta_j x + \delta_j y \right)$$

$$S_k = \frac{1}{2A} \left(\alpha_k + \beta_k x + \delta_k y \right),$$

where A = area of the triangular element which is computed from

$$2A = x_i \left(y_j - y_k \right) + x_j \left(y_k - y_i \right) + x_k \left(y_i - y_j \right)$$

and

$$\alpha_i = x_j y_k - x_k y_j; \; \alpha_j = x_k y_i - x_i y_k; \; \alpha_k = x_i y_j - x_j y_i$$
$$\beta_i = y_j - y_k; \; \beta_j = y_k - y_i; \; \beta_k = y_i - y_j$$
$$\delta_i = x_j - x_k; \; \delta_j = x_k - x_i; \; \delta_k = x_i - x_j$$

using the element temperature distribution function into governing differential equations and minimizing them will give the required FEM.

Rectangular Element

A rectangular element is defined by four nodes as shown in Fig. 7.3, which can be used to model plane solids with rectangular geometries. The four vertices of a rectangle are the nodes of the element. The shape functions can be found using the same procedures discussed above to find shape functions for the triangular element.

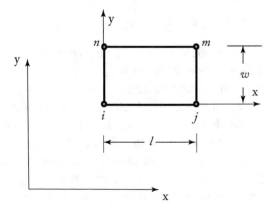

Figure 7.3: Four-node rectangular element.

Let us consider that at each node, the dependent variable is temperature. We can represent the temperature distribution function over the rectangular element, $T^{(e)}$ as

$$T^{(e)} = b_1 + b_2 x + b_3 y + b_4 xy,$$

where b_1, b_2, b_3, and b_4 are constants.

Using the following conditions:

$$@x = x_i \text{ and } y = y_i, \, T = T_i$$
$$@x = x_j \text{ and } y = y_j, \, T = T_j$$
$$@x = x_m \text{ and } y = y_m, \, T = T_m$$
$$@x = x_n \text{ and } y = y_n, \, T = T_n$$

the constants b_1, b_2, b_3, and b_4 can be determined. After finding the constants and substituting back into element temperature distribution equation, we can simplify as

$$T^{(e)} = \begin{bmatrix} S_i & S_j & S_m & S_n \end{bmatrix} \begin{Bmatrix} T_i \\ T_j \\ T_m \\ T_n \end{Bmatrix},$$

where the shape functions are given by

$$S_i = \left(1 - \frac{x}{l}\right)\left(1 - \frac{y}{w}\right); \; S_j = \frac{x}{l}\left(1 - \frac{y}{w}\right); \; S_m = \frac{xy}{lw}; \; S_n = \frac{y}{w}\left(1 - \frac{x}{l}\right).$$

The above equation represents the variation of temperature distribution (in other words, any dependent variable) over a rectangular region in terms of its nodal values.

7.3 ISOPARAMETRIC FORMULATION

In engineering, isoparametric finite element formulation is one of the most commonly used approaches. In isoparametric formulation, both shape (geometry) and field variables (nodal variables) can be described by the same shape functions. The physical element is a general quadrilateral shape. In order to cater to different sizes/shapes, a master (reference) element is used, as shown in Fig. 7.4. The physical element in local coordinates (x, y) is mapped into a reference element defined in natural coordinates (ξ, η) with reference at the center.

For example, consider a rectangular element with local coordinates (x, y) that is mapped into natural coordinates (ξ, η) with limits -1 to $+1$, as shown in Fig. 7.5. The coordinate relationship between local coordinates and natural coordinates is related as

$$\xi = \frac{2x}{l} - 1, \quad \eta = \frac{2y}{w} - 1.$$

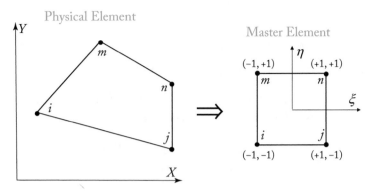

Figure 7.4: A quadrilateral element in X-Y-coordinates (left) transformed into natural coordinates.

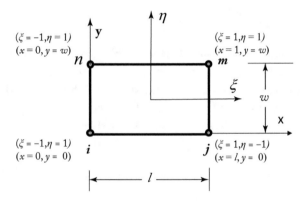

Figure 7.5: Quadrilateral element in natural coordinates and local coordinates.

The shape functions in terms of natural coordinates are

$$S_i = \frac{1}{4}(1 - \xi)(1 - \eta)$$

$$S_j = \frac{1}{4}(1 + \xi)(1 - \eta)$$

$$S_m = \frac{1}{4}(1 + \xi)(1 + \eta)$$

$$S_n = \frac{1}{4}(1 - \xi)(1 + \eta).$$

The quadrilateral element shown in Fig. 7.5 can be used in solid mechanics problems, say plane stress analysis. The displacement field (u and v) which is defined by displacement in two directions (u in x-direction, and v in y-direction), can be expressed in terms of its nodal values as

$$u(x, y) = S_i u_{ix} + S_j u_{jx} + S_m u_{mx} + S_n u_{nx}$$
$$v(x, y) = S_i v_{ix} + S_j v_{jx} + S_m v_{mx} + S_n v_{nx},$$

where S_i, S_j, S_m, and S_n are the shape functions. Now to define the geometry to represent any point within an element, we will use the same shape functions described as

$$x = S_i x_i + S_j x_j + S_m x_m + S_n x_n$$
$$y = S_i y_i + S_j y_j + S_m y_m + S_n y_n.$$

The above mapping of displacements as well as interpolating geometry using the same shape functions is called isoparametric mapping and can be extended to three dimensions as well.

7.4   HIGHER-ORDER ELEMENTS

An element with interpolation polynomial (shape functions) of the order of two or more is usually called a higher-order element. In higher-order elements, in addition to primary (corner) nodes, some secondary nodes (interior and/or mid-side) are introduced and match the nodal degrees of freedom match the number of constants in the interpolation polynomial. This concept of higher-order elements can be extended to all 1D, 2D and 3D elements. Examples of higher-order elements are presented in Fig. 7.6.

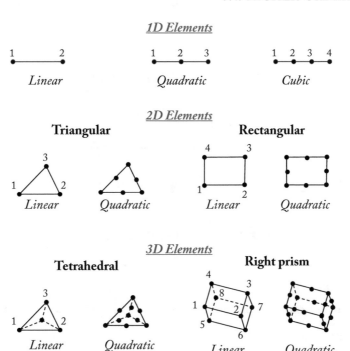

Figure 7.6: Higher-order elements in 1D, 2D, and 3D.

7.4.1   SHAPE FUNCTIONS FOR HIGHER-ORDER ELEMENTS

The shape functions for 1D higher order elements are given in Fig. 7.7.

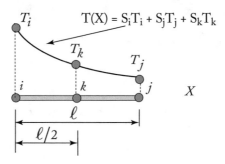

Figure 7.7: Three-node quadratic element in 1D.

The shape functions for 1D three-node Quadratic Element (Global Coordinate, X) are given below.

$$T(X) = S_i T_i + S_j T_j + S_k T_k$$

$$S_i = \frac{2}{\ell^2}(X - X_j)(X - X_k)$$

$$S_j = \frac{2}{\ell^2}(X - X_i)(X - X_k)$$

$$S_k = \frac{-4}{\ell^2}(X - X_i)(X - X_j).$$

The shape functions in terms of natural coordinate (ξ) are

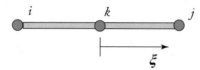

Figure 7.8: Three-node isoparametric quadratic element in 1D.

$$S_i = \frac{1}{2}\xi(1 - \xi), \quad S_j = \frac{1}{2}\xi(1 + \xi), \quad S_k = (1 + \xi)(1 - \xi).$$

1D four-node cubic element in natural coordinate (ξ).

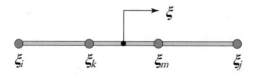

Figure 7.9: Four-node isoparametric cubic element in 1D.

$$\xi_i = -1, \quad \xi_k = -\frac{1}{6}, \quad \xi_m = \frac{1}{6}, \quad \xi_j = 1$$

$$S_i = \frac{1}{16}(1 - \xi)(3\xi + 1), \quad S_j = \frac{1}{16}(1 + \xi)(3\xi + 1)(3\xi - 1)$$

$$S_k = \frac{9}{16}(1 + \xi)(\xi - 1)(3\xi - 1), \quad S_m = \frac{9}{16}(1 + \xi)(1 - \xi)(3\xi + 1).$$

Similarly, we can find the shape functions for 2D elements. For example, the shape functions for a 2D quadratic isoparametric element are given in Fig. 7.10.

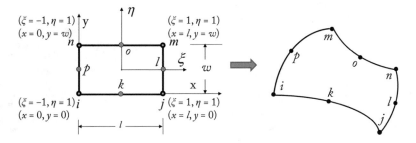

Figure 7.10: Eight-node isoparametric quadratic element in 2D.

The geometry is defined by

$$X^{(e)} = S_i X_i + S_j X_j + S_m X_m + S_n X_n + S_k X_k + S_l X_l + S_o X_o + S_p X_p$$
$$Y^{(e)} = S_i Y_i + S_j Y_j + S_m Y_m + S_n Y_n + S_k Y_k + S_l Y_l + S_o Y_o + S_p Y_p.$$

The element temperature distribution, for example, is defined by

$$T^{(e)} = S_i T_i + S_j T_j + S_m T_m + S_n T_m + S_k T_k + S_l T_l + S_o T_o + S_p T_p,$$

where

$$S_i = \frac{1}{4}(1 - \xi)(1 - \eta)(1 + \xi + \eta), \quad S_j = \frac{1}{4}(1 + \xi)(1 - \eta)(-1 + \xi - \eta)$$

$$S_m = \frac{1}{4}(1 + \xi)(1 + \eta)(-1 + \xi + \eta), \quad S_n = \frac{1}{4}(1 - \xi)(1 + \eta)(1 + \xi - \eta)$$

$$S_k = \frac{1}{2}(1 - \eta)(1 - \xi^2), \quad S_l = \frac{1}{2}(1 + \xi)(1 - \eta^2)$$

$$S_o = \frac{1}{2}(1 + \eta)(1 - \xi^2), \quad S_p = \frac{1}{2}(1 - \xi)(1 - \eta^2).$$

7.5   AXISYMMETRIC ELEMENTS

In some 3D problems, where the geometry and loading are symmetrical about an axis, the problem can be analyzed using 2D axi-symmetric finite elements. Figure 7.11 shows an axi-symmetric problem which can be analyzed using either triangular or rectangular elements. More details can be found in references given in the bibliography.

Finite Element Model

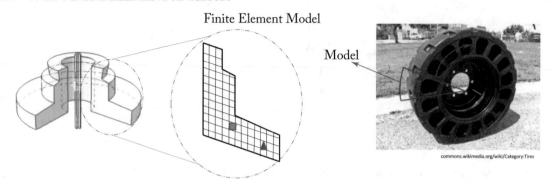

Model

commons.wikimedia.org/wiki/Category:Tires

Figure 7.11: Axisymmetric element modeling.

7.5.1   AXISYMMETRIC RECTANGULAR ELEMENT

An axisymmetric rectangular element is shown in Fig. 7.12. The shape functions are given as

$$S_i = \left(\frac{R_j - r}{l}\right)\left(\frac{Z_n - z}{w}\right), \quad S_j = \left(\frac{r - R_i}{l}\right)\left(\frac{Z_n - z}{w}\right),$$
$$S_m = \left(\frac{r - R_i}{l}\right)\left(\frac{z - Z_i}{w}\right), \quad S_n = \left(\frac{R_j - r}{l}\right)\left(\frac{z - Z_i}{w}\right).$$

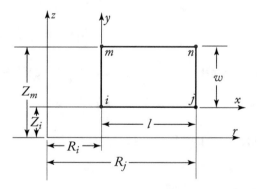

Figure 7.12: An axisymmetric rectangular element.

The above shape functions can be obtained by substituting $x = r - R_i$ and $y = z - Z_i$ in the standard rectangular element shape functions. Similarly, the shape functions can be obtained for other elements including an axisymmetric triangular element.

7.6   BASICS OF 2D STRESS ANALYSIS (PLANE SOLIDS)

Engineering structures and machine components may be approximated as 2D solids (plane solids or 2D) due to one of the dimensions (usually thickness) being either very small or very large. Examples are plane solids including a thin plate subjected to in-plane forces in both coordinate directions and a very thick solid with constant cross-section (a dam), as shown in Fig. 7.13. In this section, we will discuss how to solve 2D engineering problems for stress analysis.

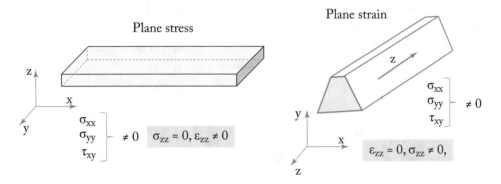

Figure 7.13: Examples of plane stress (left) and plane strain (right).

2D stress analysis problems are expressed by a system of second order partial equations, usually coupled. Based on the thickness of the plane solid, usually 2D problems can be considered as either plane stress or plane strain. The governing equations are the same for both plane stress and plane strain problems but their elastic constants matrix is different. From the stress analysis, the displacements are obtained from which strains/stresses can be obtained from the derivatives of displacements. The matrix systems of equations, similar to 1D problem are obtained using the minimum potential energy formulation when the structure/solid is in equilibrium.

The governing equations relating the 2D stresses and the displacements (u, v) in two co-ordinate (x, y) directions when a solid body (with density, r) is subjected to body forces f_x and f_y along x- and y-directions, are given below. The right-hand side term (acceleration) is zero for static problems.

Governing Differential Equation:

$$\frac{\partial \sigma_x}{\partial x} + \frac{\partial \tau_{xy}}{\partial y} + f_x = \rho \frac{\partial^2 u}{\partial t^2}$$

$$\frac{\partial \tau_{xy}}{\partial x} + \frac{\partial \sigma_y}{\partial y} + f_y = \rho \frac{\partial^2 v}{\partial t^2}.$$

Constitutive (Stress-Strain) Relations:

For a linear elastic isotropic 2D solid, the stress-strain relations are given by Hooke's law as

$$\sigma = [C]_{3\times3}\varepsilon,$$

where $\{\sigma\}$ is the stress vector, $\{\varepsilon\}$ is the strain vector, and the matrix $[C]$ is the matrix of elastic constants (E = Young's modulus and v = Poisson's ratio).

Plane stress: When the thickness direction (t) is much smaller than the width and length dimensions of a solid, then plane stress conditions exist (see Fig. 7.13). Due to thickness being small, the only non-zero stresses are: σ_{xx}, σ_{yy}, and τ_{xy}. As a result of no thickness stress ($\sigma_{zz} = 0$), the strain (ε_{zz})in the thickness direction (z) is calculated.

$$\left\{\begin{array}{c} \sigma_x \\ \sigma_y \\ \tau_{xy} \end{array}\right\} = \frac{E}{(1-v^2)} \begin{bmatrix} 1 & v & 0 \\ v & 1 & 0 \\ 0 & 0 & \frac{1-v}{2} \end{bmatrix} \left\{\begin{array}{c} \varepsilon_y \\ \varepsilon_x \\ \lambda_{xy} \end{array}\right\}$$

$$\varepsilon_{zz} = \frac{v}{(E)}\left(\sigma_{xx} + \sigma_{yy}\right).$$

Plane strain: When the thickness direction (t) is much larger than the width and length dimensions of a solid, then plane strain conditions exist. Due to thickness being large, it acts as a constraint leading to strain being zero in the thickness direction ($\varepsilon_{zz} = 0$). The only non-zero stresses are: σ_{xx}, σ_{yy}, and τ_{xy}. As a result of no thickness strain ($\varepsilon_{zz} = 0$), the stress (σ_{zz}) in the thickness direction (z) is calculated:

$$\left\{\begin{array}{c} \sigma_x \\ \sigma_y \\ \tau_{xy} \end{array}\right\} = \frac{E}{(1+v)(1-2v)} \begin{bmatrix} 1-v & v & 0 \\ v & 1-v & 0 \\ 0 & 0 & \frac{1-v}{2} \end{bmatrix} \left\{\begin{array}{c} \varepsilon_x \\ \varepsilon_y \\ \gamma_{xy} \end{array}\right\}$$

$$\sigma_{zz} = \frac{E}{(1+v)(1-2v)}\left(\varepsilon_{xx} + \varepsilon_{yy}\right).$$

Strain—Displacement Relations:

$$\varepsilon_{xx} = \frac{\partial u}{\partial x}; \quad \varepsilon_{yy} = \frac{\partial v}{\partial y}; \quad \gamma_{xy} = \frac{\partial u}{\partial y} + \frac{\partial v}{\partial x},$$

where u and v are the displacements in x-direction and y-direction, respectively.

Usually, in FE formulation, the displacements (u, v) within an element are assumed to follow some polynomials (shape functions, $\{S\}$) and we substitute them into stress-strain relations to obtain a matrix $[B]$, which is called the strain-displacement matrix. Then using minimum potential energy or weighted residual formulations, the governing stiffness matrix $[K]$ and the load vector $\{F\}$ for a given distributed load "P" on the element, can be obtained as

$$[K]^e = \int\limits_{V=Area(thickness)} B^T \, [C] \, B \, dV$$

$$F = \int\limits_{A} [S]^T \, \{P\} \, dA.$$

The details of deriving the FE formulation are not presented here and the reader may review books on FEM given in the bibliography for additional information.

The overall goal of 2D stress analysis is to relate the stresses (magnitude and location) that are obtained from FEA to the applied loads and determine the failure using the criteria as, $\sigma_{\max} \leq \sigma_o$ where σ_o is the critical/maximum stress obtained from experiments.

Principal Stresses: In FEM, the computation of principal stresses is important in order to predict failure of the solid using failure theories. For a 2D stress state shown in Fig. 7.14, the two principal stresses are given as

$$\sigma_{1,2} = \frac{\sigma_{xx} + \sigma_{yy}}{2} \pm \sqrt{\left(\frac{\sigma_{xx} + \sigma_{yy}}{2}\right)^2 + \tau_{xy}^2}.$$

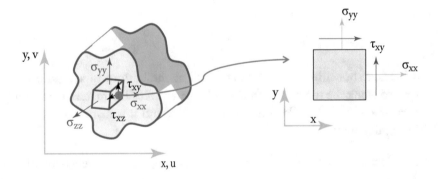

Figure 7.14: Representation of 2D stresses.

In addition to principal stresses, the maximum shear stress ($\tau_{\max}$) and von-Mises stress can be calculated as

$$\tau_{\max} = \sqrt{\left(\frac{\sigma_{xx} + \sigma_{yy}}{2}\right)^2 + \tau_{xy}^2}$$

$$\sigma_{von-Mises} = \sqrt{\sigma_1^2 - \sigma_1\sigma_2 + \sigma_2^2}.$$

Failure criteria based on stresses (the computed stresses should be within the limit of the material) and are given as

$$\sigma \leq \sigma_{yield} \quad \text{or} \quad \sigma_{ult}.$$

Safety Factor: The safety factor (FS) is defined as the ratio of yield stress to the von-Mises or maximum stress obtained from analysis. It is given as

$$FS = \frac{\sigma_{yield}}{\sigma_{max}} \quad \text{or} \quad \frac{\sigma_{ult}}{\sigma_{max}} \quad \text{or} \quad \frac{P_{max}}{P_{allowable}}.$$

7.6.1   EXAMPLES OF 2D ELEMENTS FOR STRESS ANALYSIS IN ANSYS

For a 2D plane solid stress analysis, the ANSYS software can be used to model with the following elements: PLANE 182, PLANE 183, and PLANE 13.

PLANE 182—is a four-node quadrilateral element, with two DOF (displacement in the x- and y-directions).

PLANE 183—is a eight-node quadrilateral higher order element, with two DOF (displacement in the x- and y-directions). This element is more accurate in modeling curved boundaries.

PLANE 13—is a three-node triangular element, with two DOF (displacement in the x- and y-directions).

Example 7.1
     The lift bracket shown in Fig. 7.15 is attached at the bottom to a rigid body and has a lifting force applied along the top half of the 1-in diameter hole. The material is a thin (0.075 in) steel plate ($E = 39 \times 10^6$ psi and Poisson's ratio $= 0.30$), and it is assumed that plane stress is applicable. The applied force, $F = 500$ lb.

Solution: The sequence of steps for solving the problem with ANSYS element PLANE183 is given in Table 7.2.

Table 7.1: Plane183

Element	Shape	DOF
2D Stress: PLANE183 Structural Solid, 2D-8 Node		UX, UY

Table 7.2: ANSYS steps for 2D stress analysis *(Continues.)*

I. Pre-Processing	Input
1. Specify element type	
a. Define Element Type: Preprocessor → Element Type → Add/Edit/Delete → Add	Structural Mass, Solid, 8 node 183
b. Define Element Options: Preprocessor → Element Type → Add/Edit/Delete → Options	Element shape <Triangle>, Element behavior <Plane strs w/ thk>, Element formulation <Pure displacement>
2. Specify real constants:	
a. Area: Preprocessor → Real Constants → Add/Edit/ Delete → Add	THK <Plate thickness>
3. Define material properties:	
a. Preprocessor → Material Props → Material Models → Material Model Number 1	Linear, Elastic, Isotropic, EX <Young's modulus>, PRXY <Poisson's ratio>
4. Create model geometry	
a. Create Keypoints: Preprocessor → Modeling → Create → Keypoints → In Active CS	NPT <Keypoint number>, X,Y, Z <Location of keypoint in active coordinate system>
b. Create area (solid): Preprocessor → Modeling → Create → Areas → Arbitrary → Through KP's	Select keypoints
c. Create area (hole): Preprocessor → Modeling → Create → Areas → Circle → Solid Circle	Enter coordinates of circle center and radius
d. Subtract circular area: Preprocessor → Modeling → Operate → Booleans → Subtract → Areas	Subtract circular area from solid area
e. Mesh model: Preprocessor → Meshing → MeshTool	Check Smart Size, Fine mesh (e.g., 2), Mesh: Areas, Shape: Tri, Free → Mesh → Pick All
5. Apply boundary conditions	
a. Define pressure load: Preprocessor → Loads → Define Loads → Apply → Structural → Pressure → On Lines	Select upper two quadrants of hole, Along vertical direction DOF <0>; apply pressure calculated from force and area
b. Define displacements: Preprocessor → Loads → Define Loads → Apply → Structural → Displacement → On Lines	Select bottom edge and apply displacements (both u and v) to be zero.

Table 7.2: *(Continued.)* ANSYS steps for 2D stress analysis

II. SOLUTION: Solution → Solve → Current LS	
III. POST-PROCESSING	
1. Plot results	
a.Deformed shape: General Postprocessing → Plot Results → Deformed Shape	
b. Plot Displacement: General Postprocessing → Plot Results → Contour Plot → Nodal Solution	DOF Solution, Rotation vector sum
c. Plot Stress Distribution: General Postprocessing → Plot Results → Contour Plot → Element Solution	Element Solution, Stress, vonMises stress
2. List results	
a. Nodal solution: General Postprocessing → List Results → Nodal Solution	DOF Solution, Displacement vector sum
b. Element solution: General Postprocessing → List Results → Element Solution	Line Element Results, Element Results

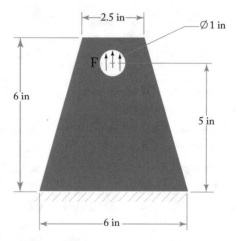

Figure 7.15: Plane stress analysis example.

After defining the element type, click on options to specify the element options (Element shape: triangle; Element behavior: plane strs w/thk; Element formulation: pure displacement). The plate thickness is entered in Real Constants. To create the geometry, use keypoints to define the corners, and create an area from the keypoints. The hole area can be created using the circle tool, entering the coordinates of the circle {3, 5}, and then subtracting the circular area from the larger area.

The use of 2D elements requires the user to specify the finite element mesh. Using the Smart Mesh feature in the Mesh Tools, the user can easily control the coarseness of the initial mesh with the slider. The mesh parameters used in this example are shown in Fig. 7.16.

After the mesh parameters have been entered, click "Mesh," then "Pick All" to mesh the model. The meshed model should appear similar to the screen shot, as in Fig. 7.17.

For the boundary conditions, apply zero displacement (all DOF) for the lines representing the upper two quadrants of the hole, and zero displacement in the x direction along the bottom edge. Apply a pressure of -1111 lb/in² (500 lb/area of applied load). Make sure the pressure is negative so that it acts in the direction of the outward normal of the surface. The results for displacement and stress distribution screen shots are given in Fig. 7.18.

7.7   BASICS OF 2D HEAT TRANSFER ANALYSIS

For a 2D steady-state heat transfer problem, the governing differential equation is given by

$$k_x \frac{\partial^2 T}{\partial x^2} + k_y \frac{\partial^2 T}{\partial Y^2} + \dot{q} = 0,$$

where k_x and k_y are thermal conductivities in x- and y-directions, $\dot{q}$ is the heat source strength, and T is unknown temperature variation over the 2D surface. The above equation along with boundary conditions can be solved for temperature distribution within a 2D domain.

7.7.1   RECTANGULAR ELEMENT

Consider a 2D bilinear rectangle element for heat transfer analysis, as shown in Fig. 7.19. The nodal variable is the temperature.

Therefore, the temperature distribution within the rectangular element is given by

$$T^{(e)} = b_1 + b_2 x + b_3 y + b_4 xy.$$

The above equation can be written as

$$T^{(e)} = \begin{bmatrix} S_i & S_j & S_m & S_n \end{bmatrix} \begin{Bmatrix} T_i \\ T_j \\ T_m \\ T_n \end{Bmatrix},$$

Figure 7.16: Mesh parameters.

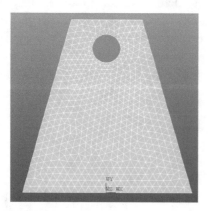

Figure 7.17: Meshed model.

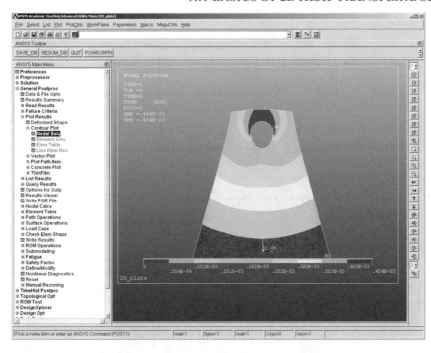

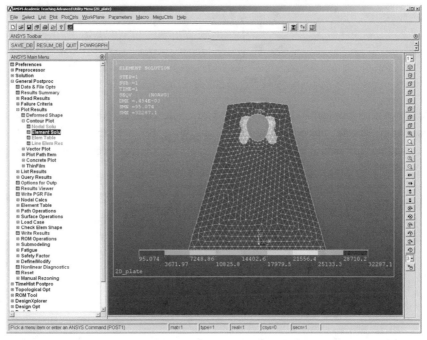

Figure 7.18: Results for displacement and stress.

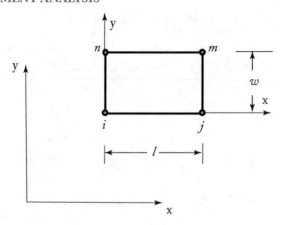

Figure 7.19: Four-node rectangular finite element.

where S_i, S_j, S_m, S_n are the shape functions and T_i, T_j, T_m, T_n are the nodal temperatures. The shape functions in local coordinates (x, y) are given by

$$S_i = \left(1 - \frac{x}{l}\right)\left(1 - \frac{y}{w}\right)$$

$$S_j = \frac{x}{l}\left(1 - \frac{y}{w}\right)$$

$$S_m = \frac{xy}{lw}$$

$$S_n = \frac{y}{w}\left(1 - \frac{x}{l}\right).$$

The element matrices are derived by substituting the assumed solution for temperature (T^e) into the governing differential equation and simplifying the equations will result in $[K]\{T\} = \{F\}$. The elemental conductance matrix $[K]$ and the thermal-load vector $\{F\}$ is given by

$$[K]^e = \frac{k_x w}{6l}\begin{bmatrix} 2 & -2 & -1 & 1 \\ -2 & 2 & 1 & -1 \\ -1 & 1 & 2 & -2 \\ 1 & -1 & -2 & 2 \end{bmatrix} + \frac{k_y l}{6w}\begin{bmatrix} 2 & 1 & -1 & -2 \\ 1 & 2 & -2 & -1 \\ -1 & -2 & 2 & 1 \\ -2 & -1 & 1 & 2 \end{bmatrix}$$

$$\{F\}^e = \frac{\dot{q} A}{4}\begin{Bmatrix} 1 \\ 1 \\ 1 \\ 1 \end{Bmatrix},$$

where $\dot{q}$ represents the heat generation per unit volume having unit depth and "A" represents the area of the element. If, however, there is a constant heat-flux (q_0) along an edge (ij) of a rectangular element, the elemental load matrix is given by

$$\{F\}^e = \frac{q_0 l_{ij}}{2} \begin{Bmatrix} 1 \\ 1 \\ 0 \\ 0 \end{Bmatrix}.$$

The next step is to assemble the element matrices to get global matrices in the form $[K]\{T\} = \{F\}$ and solve for the unknown nodal values $\{T\}$ after applying the boundary conditions.

Example 7.2   Consider the rectangular element and the nodal value of temperatures indicated, as shown in Fig. 7.20. (a) Calculate the value of T at $x = 1$ cm and $y = 1.5$ cm and (b) calculate $\frac{\partial T}{\partial x}$ and $\frac{\partial T}{\partial y}$ at $x = 1$ cm and $y = 1.5$ cm.

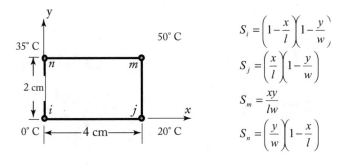

Figure 7.20: Four-node rectangular element with nodal temperatures and shape functions.

Solution:

(a)

$$T(x, y) = S_i T_i + S_j T_j + S_m T_m + S_n T_n$$

$$T(1, 1.5) = \left(1 - \frac{1}{4}\right)\left(1 - \frac{1.5}{2}\right)(0) + \left(\frac{1}{4}\right)\left(1 - \frac{1.5}{2}\right)(20)$$

$$+ \frac{(1)(1.5)}{(2)(4)}(50) + \left(\frac{1.5}{2}\right)\left(1 - \frac{1}{4}\right)(35)$$

$$T(1, 1.5) = 0 + 1.25 + 9.375 + 19.6875$$

$$T(1, 1.5) = 30.3125°C.$$

(b) Calculate $\frac{\partial T}{\partial x}$ and $\frac{\partial T}{\partial y}$ at $x = 1$ and $y = 1.5$ cm

$$\frac{\partial T}{\partial x}(x, y) = \frac{\partial S_i}{\partial x} T_i + \frac{\partial S_j}{\partial x} T_j + \frac{\partial S_m}{\partial x} T_m + \frac{\partial S_n}{\partial x} T_n$$

$$\frac{\partial T}{\partial x}(x, y) = \left(1 - \frac{1.5}{2}\right)\left(-\frac{1}{4}\right)(0) + \left(1 - \frac{1.5}{2}\right)\left(\frac{1}{4}\right)(20)$$

$$+ \frac{1.5}{(2)(4)}(50) + \frac{1.5}{2}\left(-\frac{1}{4}\right)(35)$$

$$\frac{\partial T}{\partial x}(x, y) = 0 - 1.25 + 9.375 + (-6.5625)$$

$$\frac{\partial T}{\partial x}(x, y) = 4.0625°C/m$$

$$\frac{\partial T}{\partial y}(x, y) = \frac{\partial S_i}{\partial y} T_i + \frac{\partial S_j}{\partial y} T_j + \frac{\partial S_m}{\partial y} T_m + \frac{\partial S_n}{\partial y} T_n$$

$$\frac{\partial T}{\partial y}(x, y) = \left(1 - \frac{1}{4}\right)\left(-\frac{1}{2}\right)(0) + \left(\frac{1}{4}\right)\left(-\frac{1}{2}\right)(20)$$

$$+ \frac{1}{(2)(4)}(50) + \left(1 - \frac{1}{4}\right)\left(\frac{1}{2}\right)(35)$$

$$\frac{\partial T}{\partial y}(x, y) = 0 + (-2.5) + 6.25 + 13.125$$

$$\frac{\partial T}{\partial y}(x, y) = 16.875°C/m.$$

7.7.2 EXAMPLES OF 2D ELEMENTS FOR HEAT TRANSFER ANALYSIS IN ANSYS

For a 2D steady-state heat transfer analysis, the ANSYS software can be used to model the following elements: PLANE 13, PLANE 35, and PLANE 77.

PLANE 13—is a three-node triangular element, with one DOF per node with temperature as the nodal variable.

PLANE 35—is a six-node triangular higher order element, with one DOF per node with temperature as the nodal variable. This element is more accurate in modeling curved boundaries.

PLANE 77—is an eight-node quadrilateral element for modeling 2D heat conduction problems. The element has one DOF per node with temperature as the nodal variable.

Example 7.3 The turbine blade section shown in Fig. 7.21 has internal fluid cooling channels, and is exposed to high temperature gas on the outer surfaces. Steady state conditions are assumed. The material properties are: thermal conductivity, $k = 52$ W/mK; convective coefficient, outer, $h_o = 500$ W/m^{2}K; $T_{\infty,o} = 1000°$C; and convective coefficient, channel, $h_i = 200$ W/m^{2}K; $T_{\infty,i} = 100°$C. Estimate the temperature distribution and heat flux in the blad section.

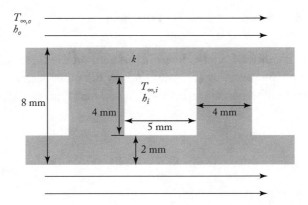

Figure 7.21: Turbine blade (shaded area) with interior cooling channels.

Solution: The ANSYS element PLANE77 will be used for the analysis, and the necessary steps are given in Table 7.4.

Due to symmetry of the problem, we can simplify the analysis model as shown in Fig. 7.22 and the three surfaces indicated can be considered adiabatic, i.e., having no heat flux through them.

The simplified model can be created using keypoints for the corners and creating an area from the keypoints. Use the Mesh Tool to generate a fine mesh with Quad shape elements on

Table 7.3: Plane77

Element	Shape	DOF
Heat transfer: PLANE77 Thermal, 2D-8 Node		Temp

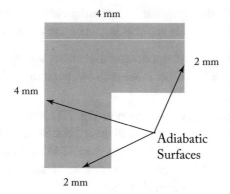

Figure 7.22: Simplified model using symmetry of the problem.

the area. For boundary conditions, apply Heat Flux of zero on the three adiabatic surfaces, and Convection on the convective surfaces ($T_{\infty,o}$ and h_o on the outer surface, $T_{\infty,i}$ and h_i on the inner surfaces). The results of temperature distribution and heat flux screen shots are presented in Fig. 7.23.

Table 7.4: ANSYS steps for 2D heat transfer analysis

I. Pre-Processing	Input
1. Specify element type:	
a. Preprocessor → Element Type → Add/Edit/Delete → Add	Thermal Mass, Solid, 8node 77
2. Specify real constants:	
NA for this type element	AREA <Cross-sectional area>
3. Define material properties:	
a. Preprocessor → Material Props → Material Models → Material Model Number 1	Thermal, Conductivity, Isotropic, KXX <Thermal conductivity>
4. Create model geometry	
a. Create Keypoints: Preprocessor → Modeling → Create → Keypoints → In Active CS	NPT <Keypoint number>, X,Y, Z <Location of keypoint in active coordinate system>
b. Create area: Preprocessor → Modeling → Create → Areas → Arbitrary → Through KP's	Select keypoints
c. Mesh model: Preprocessor → Meshing → MeshTool	Check Smart Size, Fine mesh (e.g., 2), Mesh: Areas, Shape: Tri, Free → Mesh → Pick All
5. Apply boundary conditions	
a. Define thermal loads: heat fluxes: Preprocessor → Loads → Define Loads → Apply → Thermal → Heat Flux → On Lines	Select adiabatic surfaces, VALI Heat flux <0>
b. Define thermal loads: convections: Preprocessor → Loads → Define Loads → Apply → Thermal → Convection → On Lines	Select convective surfaces and apply respective h and T values: VALI Film coefficient <convective coefficient>, VAL2I Bulk temperature < T_∞>
II. SOLUTION: Solution → Solve → Current LS	
III. POST-PROCESSING	
1. Plot results	
a. Plot temperature distribution: General Postprocessing → Plot Results → Contour Plot → Nodal Solution	
b. Plot heat flux: General Postproc → Plot Results → Vector Plot → Predefined	Item Vector to be plotted <Flux & gradient>, <Thermal flux TF>

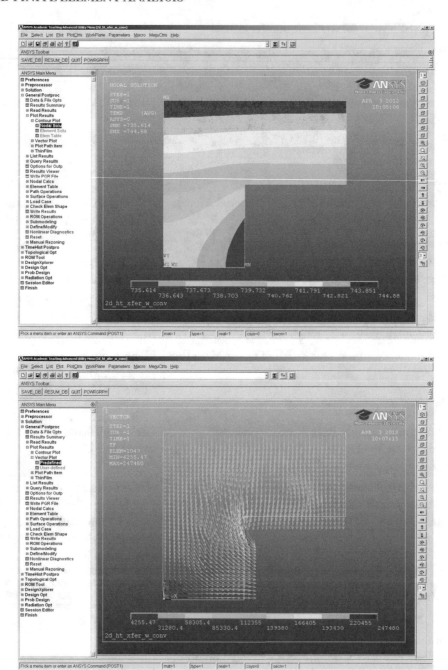

Figure 7.23: The results of temperature distribution and heat flux.

7.8   BASICS OF FLUID MECHANICS ANALYSIS

For problems involving fluid mechanics, the basic equations governing motion of fluid are the Navier-Stokes equations. These are given as:

Conservation of Mass: $\dfrac{\partial \rho}{\partial t} + \dfrac{\partial}{\partial x_k}(\rho u_k) = 0$

Conservation of Momentum: $\dfrac{\partial \rho u_t}{\partial t} + \dfrac{\partial}{\partial x_k}(\rho u_t u_k - \tau_{tk}) + \dfrac{\partial P}{\partial x_t} = S_t$

Conservation of Energy: $\dfrac{\partial(\rho E)}{\partial t} + \dfrac{\partial}{\partial x_k}((\rho E + P)u_k + q_k - \tau_{tk}u_t) = S_k u_k + Q_H,$

where

$$u = \text{fluid velocity}$$
$$Si = \text{external volume force per unit mass}$$
$$\rho = \text{fluid density}$$
$$E = \text{total energy per unit mass}$$
$$\tau_{tk} = \text{viscous shear stress tensor}$$
$$Q_H = \text{heat source per unit volume}$$
$$q_i = \text{diffusive heat flux.}$$

If we consider fluid problems of ideal flow (incompressible and inviscid), examples of which include flow around a cylinder or an airfoil or out of an orfice, etc., the 2D finite element formulation can be formulated in terms of a velocity potential function or a stream function. Then the governing differential equation will be a Laplace equation which can be solved along with boundary conditions. The details are not discussed here and the reader is referred to other FEM books included in the bibliography.

Figure 7.24 shows the velocity profile of a fully developed laminar flow of a viscous fluid between two parallel plates. The flow is driven by a difference in pressure, Δp, between the left and right ends of the channel. In this case, $p_1 > p_2$, so the fluid flows in the $+x$-direction. The "no-slip" condition is assumed to be applied at the walls of the channel, i.e., the velocity is zero at $y = 0$ and at $y = h$. The flow is also assumed to be fully-developed, so that the velocity does not change along the length of the channel.

The solution to the Navier-Stokes equations for velocity for laminar, incompressible, steady, plane flow reduces to

$$v_x = -\frac{1}{2\mu}\frac{\partial p}{\partial x}\left[\left(\frac{h}{2}\right)^2 - \left(y - \frac{h}{2}\right)^2\right].$$

This equation describes the velocity distribution as parabola shown in Fig. 7.24, with a maximum value for v_x at $y = h/2$, and minima ($v_x = 0$) at $y = 0$ and $y = h$.

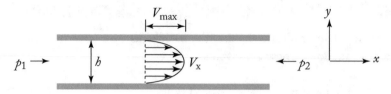

Figure 7.24: Velocity profile for incompressible, laminar flow between parallel plates.

7.8.1   EXAMPLES OF 2D ELEMENTS FOR FLUID ANALYSIS IN ANSYS

For a 2D fluid analysis, the ANSYS software can be used to model the following elements: FLUID 29, FLUID 79, and FLUID 141.

FLUID 141—is a four-node rectangular element for modeling 2D fluids within vessels with no net flow rate. The element has two DOF per node with translations in x- and y-directions at each node.

FLUID 29—is a four-node quadrilateral element, with one DOF per node with pressure as the nodal variable.

FLUID 79—is a four-node rectangular element, with one DOF per node with pressure as the nodal variable.

Example 7.4   Consider an ideal airflow around a cylinder as shown in Fig. 7.25. The properties are: air density, $\rho = 1.23$ kg/m^3; air viscosity, $\mu = 1.79 \times 10^{-5}$ N-s/m^2; freestream velocity, $U = 10$ cm/s; cylinder diameter, $D = 5$ cm; and channel width, $W = 50$ cm. Find the velocity distribution around the cylinder using ANSYS.

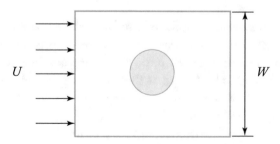

Figure 7.25: 2D flow around a cylinder problem.

Solution: The ANSYS element **FLOTRAN141** will be used for the analysis, and the necessary steps are given in Table 7.6.

Table 7.5: Flotran141

Element	Shape	DOF
Fluid: **FLOTRAN141** Fluid-Thermal, 2D-4 Node	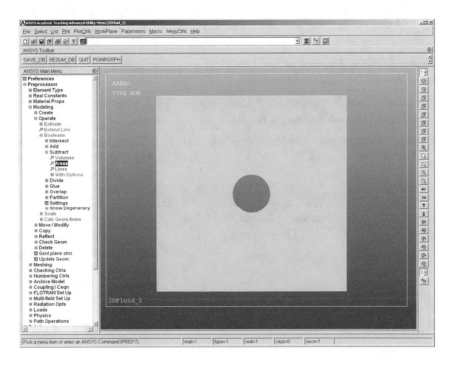	VX, VY, VZ, PRES, TEMP, ENKE, ENDS

After selecting the element type (2D FLOTRAN 141), the FLOTRAN pre-processor needs to be set-up using the menu path: Preprocessor → FLOTRAN Set Up → Fluid Properties. Select "AIR-SI" for the density and viscosity, and leave the conductivity and specific heat at the default values of "Constant." The material model is then defined as with previous examples, using the given values for the air density and viscosity.

Generate the model geometry by first creating a rectangular area to represent the channel, with one corner at the origin and 0.5 (50 cm) for the width and height. Create the circular area in the middle of the channel (center point {0.25, 0.25}), then subtract the circle from the rectangle using the Boolean operation. The model should appear as shown in the screen shot in Fig. 7.26.

Figure 7.26: Model geometry.

Table 7.6: ANSYS steps for 2D fluid flow analysis *(Continues.)*

I. Pre-Processing	Input
1. Specify element type:	
a. Preprocessor → Element Type → Add/Edit/Delete → Add	FLOTRAN CFD, 2D FLOTRAN 141
2. FLOTRAN Setup:	
a. Preprocessor → FLOTRAN Set Up → Fluid Properties	Density <AIR-SI>, Viscosity <AIR-SI>, Conductivity <Constant>, Specific heat <Constant>
3. Define material properties:	
a. Preprocessor → Material Props → Material Models → Material Model Number 1	CFD, Density, DENS <Air density>; CFD, Viscosity, VISC <Air viscosity>
4. Create model geometry	
a. Create area (channel): Preprocessor → Modeling → Create → Areas → Rectangle → By 2 Corners	NPT <Keypoint number>, X,Y, Z <Location of keypoint in active coordinate system>
b. Create area (circle): Preprocessor → Modeling → Create → Areas → Circle → Solid Circle	Enter center point and radius of circle
c. Subtract circular area: Preprocessor → Modeling → Operate → Booleans → Subtract à Areas	Subtract circular area from solid area
d. Mesh model: Preprocessor → Meshing → MeshTool	Check Smart Size, Fine mesh (e.g., 2), Mesh: Areas, Shape: Tri, Free à Mesh à Pick All
5. Apply boundary conditions	
a. Define velocities: Preprocessor → Loads → Define Loads → Apply → Fluid/CFD → Velocity → On Lines	Select edges with zero velocity (upper and lower edges and circular edge) and apply velocity of "0". Apply a velocity of "0.1" (10cm/s) to the inlet and outlet edges.
b. Define pressures: Preprocessor → Loads → Define Loads → Apply → Fluid/CFD → Pressure DOF → On Lines	Select inlet and outlet edges and apply zero pressure (atmospheric) to each

Table 7.6: *(Continued.)* ANSYS steps for 2D fluid flow analysis

II. SOLUTION: Solution → Solve → Current LS	
III. POST-PROCESSING	
1. Plot results	
a. Fluid velocity (contour plot): General Postproc → Plot Results → Contour Plot → Nodal Solution	DOF Solution, Fluid velocity
b. Fluid velocity (vector plot): General Postproc → Plot Results → Vector Plot → Predefined	Item Vector to be plotted <DOF Solution>, <Velocity V>

A triangular mesh using the Mesh Tool is generated using a Smart Size value of 1 or 2 (Fine), the mesh and is applied to the shaded area. The screen shot of meshed model should appear as shown in Fig. 7.27.

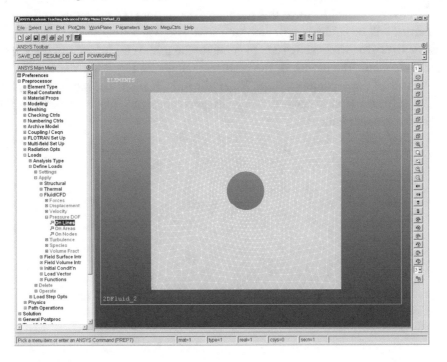

Figure 7.27: Meshed model.

The boundary conditions for the fluid analysis will be based on velocities and pressures. For this case, the "no-slip" condition (zero fluid velocity) will be assumed to apply at the walls of the channel (top and bottom edges) and at the cylinder surface. Use the menu path: Preprocessor → Loads → Define Loads → Apply → Fluid/CFD → Velocity → On Lines; select the lines at the top and bottom, and the circle, then enter "0" as the *VX* Load value. To apply the freestream velocity, repeat the sequence, selecting the inlet and outlet edges (lines on the left and right of the rectangle), and enter a *VX* Load value of "0.10" (10 cm/s).

Run the solution and generate the desired results in Post-Processing. The results of the velocity contour and vector plots obtained from the analysis are shown in screen shots in Fig. 7.28.

Note: In addition to the 2D elements discussed in this chapter, ANSYS offers many more elements for modeling and the reader is requested to refer to the ANSYS manual for more element capabilities and descriptions.

Example 7.5   Diffusion-based microfluidic drug delivery. A diffusion based microfluidic device is shown in Fig. 7.29. Find the rate of mass at outlet.

Solution: The drug delivery from the reservoir through the channels is modeled by a diffusion equation. Since the underlying principle of the diffusion equation is similar to the heat equation, we will use ANSYS thermal to solve this problem. The notions Temperature and Heat Flux in ANSYS Thermal will be equivalent to Concentration and Mass Flux in microfluidic problems. We will use ANSYS Thermal steady state to generate the initial condition, and then run the case in transient component. The model for the problem is shown in Fig. 7.29. We want to make sure the model consists of two separate sections: Reservoir and Channel. The model can be built using external CAD software. If so, it needs to be exported to compatible file format such as IGES and Parasolid Binary (recommended).

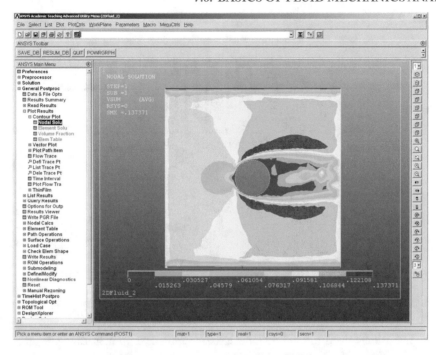

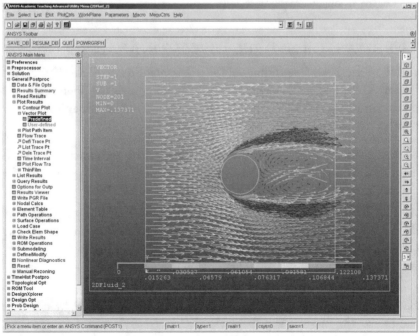

Figure 7.28: Results of the velocity contour and vector plots.

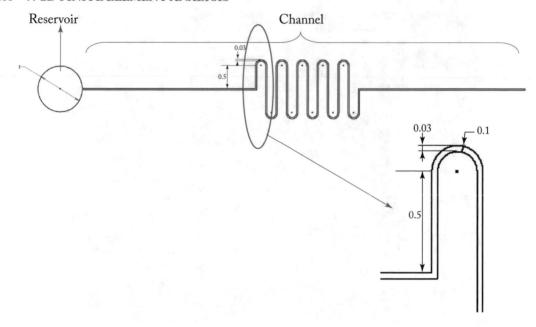

Figure 7.29: Diffusion-driven microfluidic device. Reservoir is 1 mm in diameter, the channel is 0.03 mm in width, the whole channel length is 10 mm.

7.8.2 ANSYS STEPS FOR 2D DRUG DIFFUSION ANALYSIS

1. Open ANSYS workbench and drag ANSYS Thermal Steady State (Fig. 7.30a). Click on Engineering Data, and make new material by naming it in the appropriate field. Set the value as shown in Fig. 7.30b. Right-click on steady state at line Solution → Transfer → to new transient. Now we have two connected solvers (Fig. 7.30c).

   Right-click on Geometry section of Steady-State Thermal, and choose "Import Geometry." Import.

2. Double-click the ANSYS steady-state Thermal Model section (line 4) to open the GUI for ANSYS Mechanical for thermal case. We have two analyses in the GUI, as shown in Fig. 7.31a: Steady-State Thermal and Transient-Thermal, as highlighted in Fig. 7.31a. Set the unit for temperature as *K* from Units menu (Fig. 7.31b).

3. Select Mesh. Choose "sizing" set-up as shown in Fig. 7.32a. Then right-click mesh, choose generate mesh. The mesh is shown in Fig. 7.32b. ANSYS will determine the optimum arrangement of mesh, and will use different type of mesh given the nature of geometry of the model.

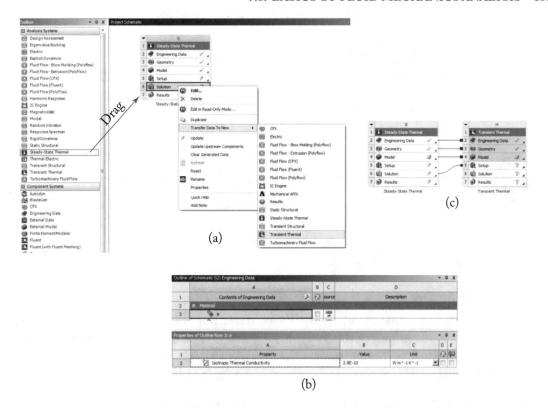

Figure 7.30: (a) and (c) preparing the analysis, (b) preparing material properties.

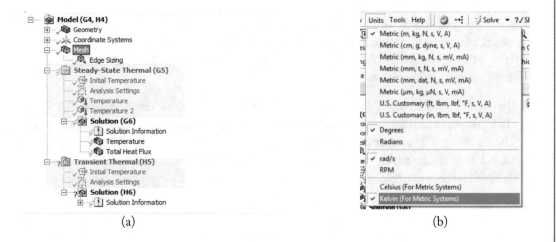

Figure 7.31: (a) The Workbench has two analyses: the Steady-State and Transient; (b) setting up unit.

Sizing	
Use Advanced Si...	On: Fixed
Relevance Center	Medium
Initial Size Seed	Active Assembly
Smoothing	High
Min Size	1.e-005 m
Max Face Size	3.e-005 m
Growth Rate	Default
Minimum Edge L...	2.2505e-007 m

(a)

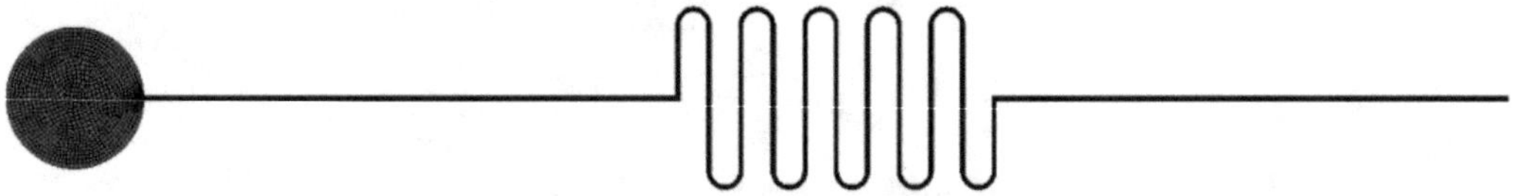

(b)

Figure 7.32: (a) Mesh set-up, (b) Meshed model.

4. Set the "Initial Temperature" as 0. Right-click at Steady-State Thermal, and choose Temperature to set up boundary conditions. We need two Temperature boundary condition. The first is for the whole body of Reservoir, and the second is on the edge of outlet (Fig. 7.33b). Set the magnitude of Temperature in the reservoir at 5,000. Finally, right-click Solution, choose Solve.

5. The initial concentration of the problem is generated. Click on Initial Temperature under Transient Thermal, and make sure the Initial Temperature Environment shows "Steady-State Thermal."

6. Click analysis setting and set the end time, as shown in Fig. 7.34(c). The end time should be set according to problem. In this case, it needs days for the drug to diffuse. In this case, it is better to choose the Auto Time Stepping controlled by ANSYS. Then, right-click on solution and choose Solve.

7. To explore the solution, right-click Solution, choose Thermal → Temperature (Fig. 7.35a). The Temperature represents the concentration in microfluidic case. Choose "Graph" tab as shown in Fig. 7.35b to see concentration at specific time step, or generate animation. In this case, it is found that after about 11 days, approximately 32% of drug from the Reservoir has diffused but has not reached the outlet (Fig. 7.35c). Hence, a longer time period is needed.

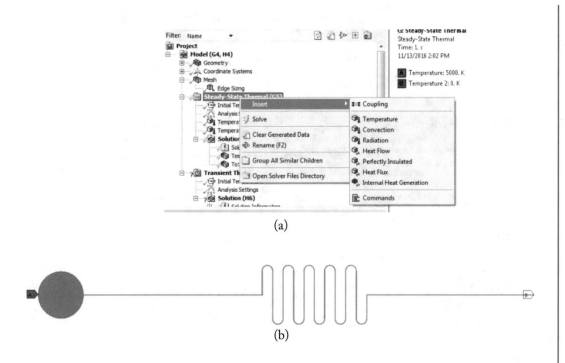

(a)

(b)

Figure 7.33: (a) Boundary condition set-up; (b) the location of boundary condition.

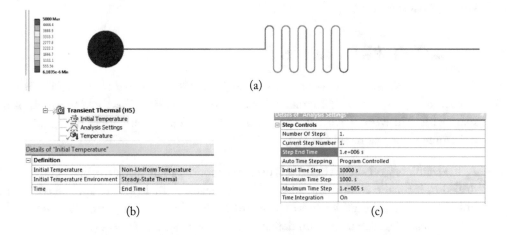

(a)

(b)   (c)

Figure 7.34: (a) The steady state result is used for initial condition in Transient Analysis, (b) the Initial Temperature section in Transient Thermal automatically imports Steady State result, and (c) time step set-up for this problem.

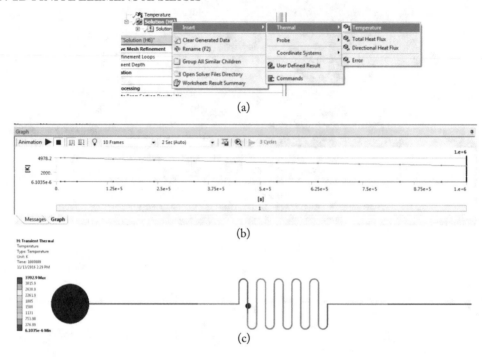

(a)

(b)

(c)

Figure 7.35: (a) Generate concentration distribution to visualize computation result, (b) the Graph bar and tab to choose result at specific time step, and (c) concentration distribution after 1×10^6 s (approximately 11 days).

7.9   EXERCISE PROBLEMS

7.1.   A 304 stainless steel cantilever bar has a weight of 120 N applied to its tip. The bar has a thickness of 5 cm and a Modulus of Elasticity of 190 GPa. The Yield Strength of the material is 276 MPa. Using ANSYS, calculate the Von-Mises stress distribution and determine if the beam will fail (Fig. 7.36).

7.2.   A 304 stainless steel square bracket is used to support a tight cord. The bracket has a hole drilled in it for the cord to thread through it, meaning the loads are applied within the hole. The bracket has a thickness of 2 cm and a Modulus of Elasticity of 190 GPa. Using ANSYS, calculate the maximum deflection and the maximum stress (Fig. 7.37).

7.3.   Consider a small rectangular aluminum plate with dimensions 40 cm × 10 cm and a thermal conductivity value of $k = 168$ W/m·k. The two ends of the plate have constant temperatures of 17°C each and the surrounding fluid has a temperature of 200°C. Using manual calculations, determine the temperature distribution within the plate, under

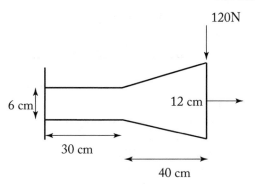

Figure 7.36: Stainless steel cantilever bar.

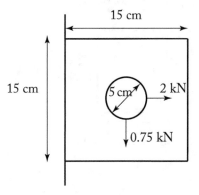

Figure 7.37: Stainless steel square bracket.

steady state conditions (Fig. 7.38). (*Hint:* Because there are two axes of symmetry, you should model only a quarter of the plate.)

7.4. A square aluminum fin is attached to a heat sink at a constant temperature of 237°C with dimensions 3 cm × 3 cm and a thermal conductivity value of $k = 168$ W/m·k. The air surrounding the fin is has a constant temperature of 20°C. The top and bottom portions of the plate have air blown over them resulting in a convection coefficient of $h = 200$ W/m²·k, while the end portion of the fin does not and free convection results in a convection coefficient of $h = 30$ W/m²·k. Using manual calculations, determine the temperature distribution within the plate, under steady state conditions (Fig. 7.39). (*Hint:* Because there is an axis of symmetry, you should model one-half of the fin.)

7.5. A truncated triangular fin is attached to a surface at a constant temperature of 50°C. Aluminum alloy 6063 is used with a thermal conductivity value of $k = 201$ W/m·k. The

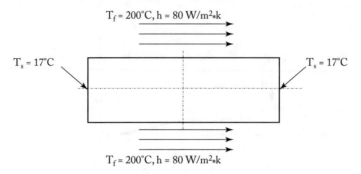

Figure 7.38: Rectangular aluminum plate (Problem 3).

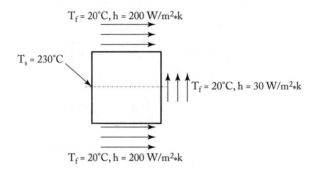

Figure 7.39: Square aluminum fin.

fin has boundary conditions given in Fig. 7.40. Using ANSYS, create the model and find the temperature distribution and heat flux within the pin for:

(a) air forced over at 20°C and $h = 200$ W/m²·k and

(b) water forced over at 20°C and $h = 7000$ W/m²·k.

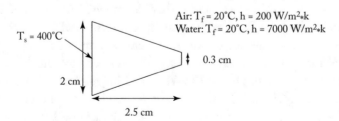

Figure 7.40: A truncated triangular fin.

7.6. In an effort to reduce enhanced heat transfer rates, an idea has been proposed that the inside surface of the pipe be extended to form triangular fins, as shown in Fig. 7.41. The brass pipe is used to cool just-boiled water. Using ANSYS, find the temperature distribution and heat flux for:

   (a) the pipe without the fins;

   (b) the pipe with the addition of the fins (*Hint:* To build the fins switch ANSYS to polar coordinates.); and

   (c) what was the effect of the fins?

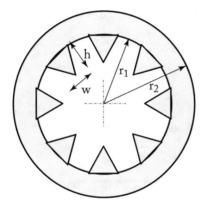

$r_1 = 10$ cm
$r_2 = 13$ cm
$w = 1$ cm
$h = 2$ cm

$k_{brass} = 109$ W/m*k
$T_{inside} = 99°C$
$h_{inside} = 2500$ W/m²*k
$T_{outside} = -8°C$
$h_{inside} = 25$ W/m²*k

Figure 7.41: Inside surface of the pipe.

7.7. Hot water flows through pipes that are embedded in a concrete slab as a method of keeping a room warm on a cold winter day. A section of the slab is shown in Fig. 7.42. The temperature inside the pipes is 70°C with a corresponding heat transfer coefficient of $h = 1000$ W/m²·k. Assume the side walls are perfectly insulated and neglect the thermal resistance through the pipe walls. Using ANSYS, find the temperature distribution of the wall.

7.8. Water is moving in a pipe that has a diffuser attached to it, as shown in Fig. 7.43. The fluid is initially moving uniformly and is at 20°C with a viscosity of 1.002 mPa·s and a density of 998.2 kg/m³. Using ANSYS, find the Fluid velocity at the end of the diffuser.

7.9. Water is moving in a pipe, as shown in Fig. 7.44. The fluid is initially moving uniformly and is at 20°C with a viscosity of 1.002 mPa·s and a density of 998.2 kg/m³. Using ANSYS, find the Fluid velocity at the end of the pipe.

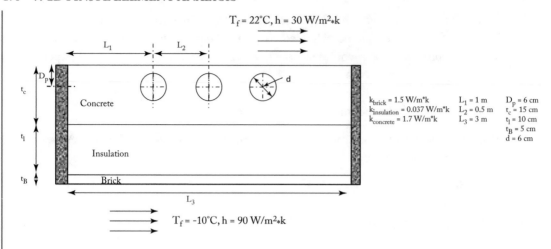

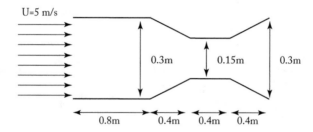

Figure 7.42: Hot water flows through pipes.

Figure 7.43: Water moving in a pipe that has a diffuser attached to it.

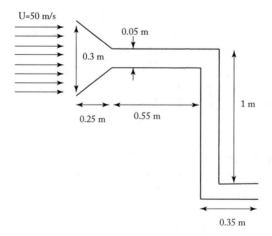

Figure 7.44: Water is moving in a pipe.

7.10. Water is moving in a pipe with a circular obstruction located in the middle of the pipe, as shown in Fig. 7.45. The fluid is initially moving uniformly and is at 20°C with a viscosity of 1.002 mPa·s and a density of 998.2 kg/m³. Using ANSYS, find the Fluid velocity at the end of the pipe.

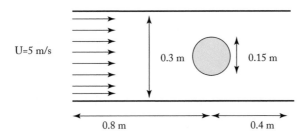

Figure 7.45: Water is moving in a pipe with a circular obstruction.

7.11. An experimental nozzle is created by having quarter circles form the tip. Water moving in the pipe is shown in Fig. 7.46. The fluid is initially moving uniformly and is at 20°C with a viscosity of 1.002 mPa·s and a density of 998.2 kg/m³. Using ANSYS, find the Fluid velocity at the end of the pipe.

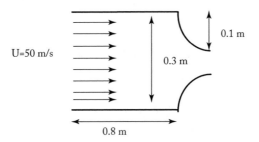

Figure 7.46: An experimental nozzle.

CHAPTER 8

3D Finite Element Analysis

After reading this chapter, you will be able to:

- understand 3D problems in engineering;

- list various types of elements for 3D problems;

- model and analyze 3D solid stress problems;

- model and analyze 3D heat transfer problems; and

- understand multiphysics problems in ANSYS.

8.1 OVERVIEW

In previous chapters, we learned how to perform 2D engineering analysis. In general, all practical/real systems/components are 3D and require modeling in three-dimensions in order to capture the behavior of the component accurately. This chapter deals with 3D elements and their application to various engineering problems including solid stress analysis and heat transfer analysis. Specifically, 3D analysis using ANSYS software is discussed.

8.2 3D ELEMENTS

In general, a solid (3D geometry) can be modeled using a variety of 3D elements, as shown in Fig. 8.1. The elements may have different shapes (rectangular brick or tetrahedral) and different nodes. The shape functions are derived using procedures similar to those discussed in previous chapters. At each node, there will be three DOF in each of the coordinate directions. These DOF depends on the type of problem we are analyzing (temperature in heat transfer analysis or displacement in stress analysis or velocity in fluid analysis and so on). The concepts from 2D elements and analysis are extended to 3D analysis.

8.3 ELEMENTS IN ANSYS FOR 3D ANALYSIS

A broad variety of 3D finite elements are available in ANSYS, and some of these elements for solid, heat transfer, and fluid flow are listed below.

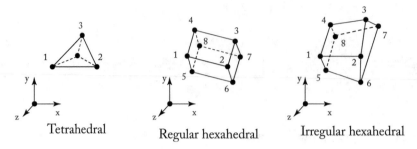

Figure 8.1: Types of 3D plane elements with different shape and number of nodes.

8.3.1   SOLID ANALYSIS ELEMENTS

For a 3D solid stress analysis, the ANSYS software can be used to model with the following elements: SOLID185, SOLID285, SOLID186, and SOLID187, as shown in Table 8.1.

SOLID 185—is an eight-node rectangular element, with three DOF (displacements u_x, u_y, and u_z in the x-, y-, and z-directions, respectively) per node.

SOLID 285—is a four-node tetrahedral element, with three DOF (displacements u_x, u_y, and u_z in the x-, y-, and z-directions, respectively) per node.

SOLID 187—is a ten-node tetrahedral higher order element, with three DOF (displacements u_x, u_y, and u_z in the x-, y-, and z-directions, respectively) per node. This element is more accurate in modeling curved boundaries.

SOLID 186—is a twenty-node quadratic higher order element, with three DOF (displacements u_x, u_y, and u_z in the x-, y-, and z-directions, respectively) per node. This element is more accurate in modeling curved boundaries.

8.3.2   HEAT TRANSFER/THERMAL ANALYSIS ELEMENTS

For a 3D steady-state heat transfer analysis, the ANSYS software can be used to model with the following elements: **SOLID** 70, **SOLID** 87, and **SOLID** 90.

SOLID 70—is an eight-node quadrilateral element for modeling 3D heat conduction problems. The element has one DOF per node with temperature as the nodal variable.

SOLID 87—is a ten-node tetrahedral higher order element, with one DOF per node with temperature as the nodal variable. This element is more accurate in modeling curved boundaries.

SOLID 90—is an twenty-node higher order quadratic element for modeling 3D heat conduction problems. The element has one DOF per node with temperature as the nodal variable.

Table 8.1: 3D solid elements for stress analysis in ANSYS

ANSYS Element	Element Shape and Nodes	Degrees of Freedom (DOF)/ Node
SOLID185, Structural Solid, 3D-8 Node		UX, UY, UZ
SOLID285, Structural Solid, 3D-4 Node		UX, UY, UZ
SOLID187, Structural Solid, 3D-10 Node		UX, UY, UZ
SOLID186, Structural Solid, 3D-20 Node		UX, UY, UZ

Table 8.2: 3D solid elements for heat transfer analysis in ANSYS

ANSYS Element	Element Shape and Nodes	Degrees of Freedom (DOF)/ Node
SOLID70, Thermal Solid, 3D-8 Node		TEMP
SOLID87, Thermal Solid, 3D-10 Node		TEMP
SOLID90, Structural Solid, 3D-20 Node		TEMP

8.3.3   FLUID ANALYSIS ELEMENTS

For a 3D fluid analysis, the ANSYS software can be used to model with the **FLOTRAN 142** elements shown in Table 8.3, which is an eight-node rectangular element, with one DOF per node with pressure as the nodal variable. In addition, higher elements can also be used.

Table 8.3: 3D solid elements for fluid analysis in ANSYS

ANSYS Element	Element Shape and Nodes	Degrees of Freedom (DOF)/ Node
FLOTRAN142, Fluid-Thermal, 3D-8 Node		VX, VY, VZ, PRES, TEMP, ENKE, ENDS

8.4   3D SOLID MODELING FOR IMPORT TO ANSYS

For conducting finite element analysis of complex 3D solid objects/geometries, first there is a need to generate the geometry. There are two ways to create the solid geometry. The first approach involves using the tools in ANSYS software and the second one can be performed using any CAD software packages (Solidworks or Inventor or Autocad). In ANSYS, one can use *keypoints* as the basis and define lines, areas, and volumes of the solid geometry. In addition, one can create 3D solid objects using volume primitives such as block, prism, cylinder, cone, sphere, and torus. Due to availability of CAD software, it is easy to create the geometry using Solidworks, and importing to ANSYS using *.iges* file extension, as shown in Fig. 8.2.

In addition to *.iges* file extension, there is another option, *.parasolid* file extension as well. After importing the file to ANSYS software, make sure the solid object's volumes/areas are clearly visible without missing any information. The following examples are provided in order to improve understanding of the steps involved in completing 3D finite element analysis with ANSYS.

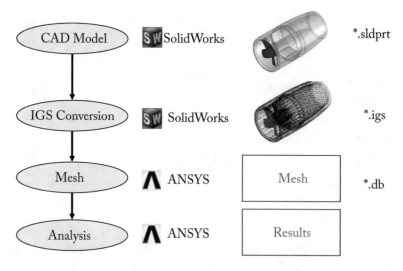

Figure 8.2: Procedure for importing SolidWorks file into ANSYS for analysis.

Example 8.1   Figure 8.3 shows a bracket to support books made from a material ($E = 73$ GPa; Poisson's ratio $= 0.30$; density $= 2780$ kg/m^3) and fixed at one end and constrained at the two holes. The load applied on the bracket is 1000 N/m^2. Find the deflection of the bracket and von-Mises stress distribution. Also conduct modal analysis to find natural frequencies and mode shapes.

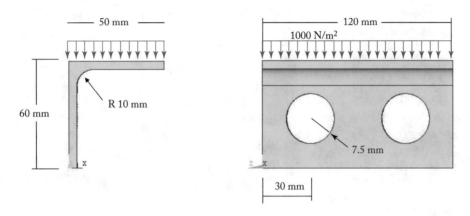

Figure 8.3: A bracket with dimensions and loading.

The following steps demonstrate how to create the model geometry and find a solution using ANSYS.

Step 1: The bracket is easily created in Solidworks, as shown Fig. 8.4.

Step 2: The Solidworks file is saved as IGES extension as it is a universal graphics format used specifically to transfer geometry from one platform to another, as shown in Fig. 8.5.

Step 3: Switch to ANSYS software and import the .igs file, as shown in Fig. 8.6 for imported bracket geometry.

Step 4: Conduct pre-processing by specifying the element type (BRICK 8 Node 185), material properties and meshing. See pre-processing activities in Figs. 8.7, 8.8, and 8.9.

Step 5: Specify constraints, loads, as shown in Fig. 8.10.

Step 6: Perform solution to find deflection and von-Mises stresses, as shown in Figs. 8.11 and 8.12.

Step 7: Perform modal solution to find natural frequencies and mode shapes by completing another analysis. For this, the density of the bracket material needs to be specified. We need to go back to the material properties screen and the density should be specified, as shown in Figs. 8.13 and 8.14.

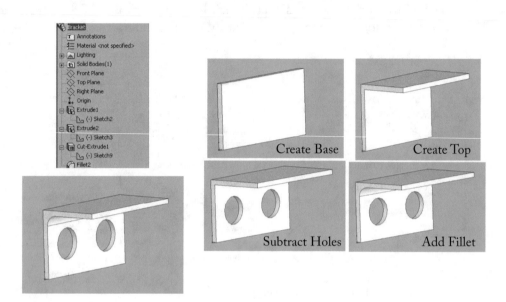

Figure 8.4: Bracket in Solidworks.

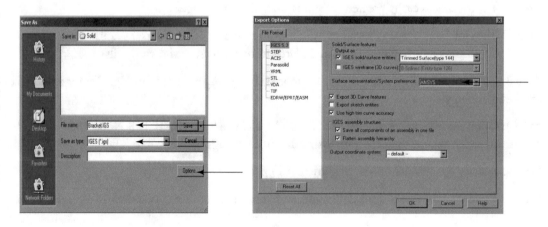

Figure 8.5: Solidworks file.

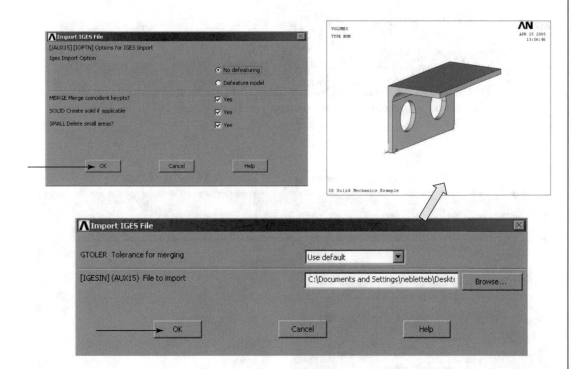

Figure 8.6: ANSYS software.

Main Menu Preprocessor → Element type → Add/Edit/Delete

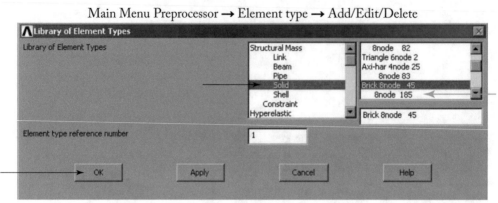

Select **Structural Mass Solid** in the left window and **Brick 8 node 185** in the right menu

Main Menu Preprocessor → Material Props → Material Models

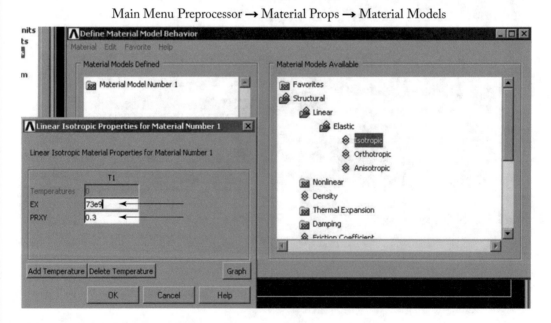

Figure 8.7: Pre-processing activities.

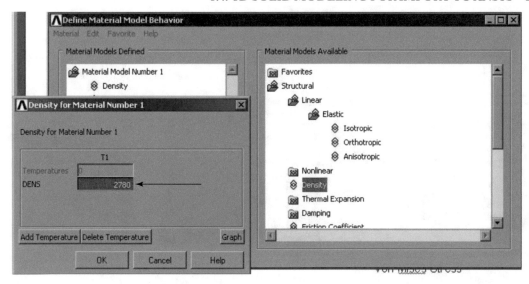

Figure 8.8: Pre-processing activities.

Main Menu **Preprocessor** → **Meshing** → **Mesh Tool**

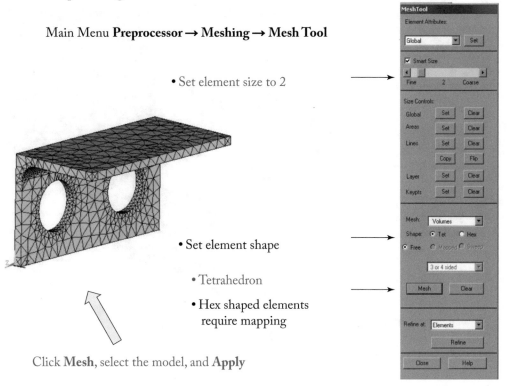

• Set element size to 2

• Set element shape

• Tetrahedron

• Hex shaped elements
  require mapping

Click **Mesh**, select the model, and **Apply**

Figure 8.9: Pre-processing activities.

Main Menu
Solution → Define Loads → Apply → Displacement → On Areas

Select the walls of the holes and **Apply**

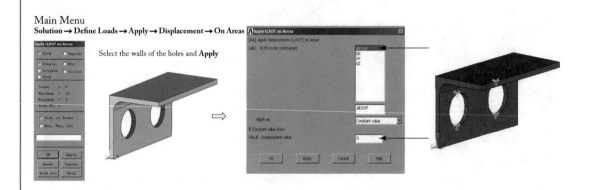

Main Menu
Solution → Define Loads → Apply → Pressure → On Areas

Select the top surface and **Apply**

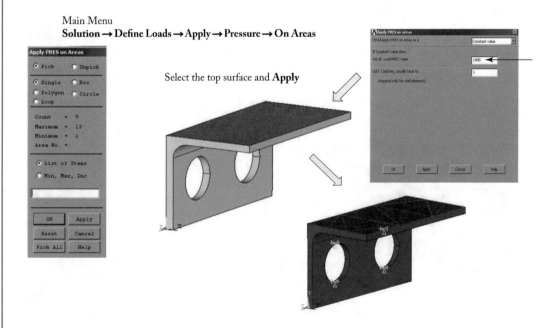

Figure 8.10: Constraints loads.

Main Menu **Solution → Solve → Current LS**

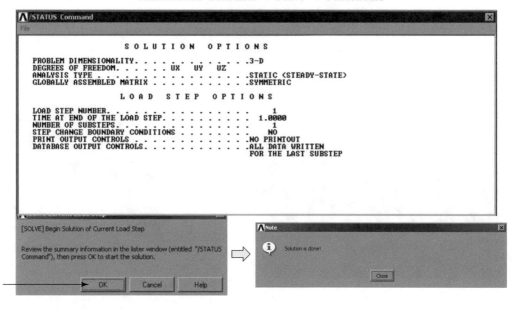

Figure 8.11: Deflection and von-Mises stresses.

Step 8: The mode shapes can be visualized by specifying the PlotCtrls, and also the animations, as shown in Fig. 8.15.

Step 9: Exit from ANSYS by saving everything, as shown in Fig. 8.16.

Main Menu
General Postproc → Plot Results → Deformed Shape

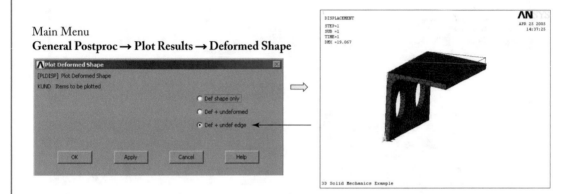

Main Menu
General Postproc → Plot Results → Contour Plot → Nodal Solu

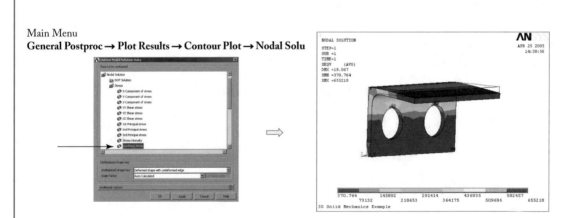

Figure 8.12: Deflection and von-Mises stresses.

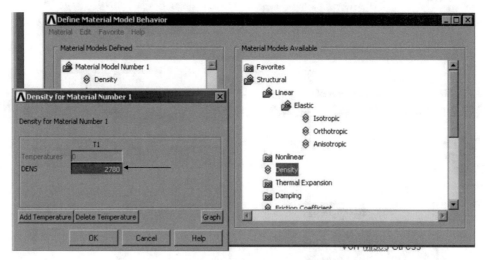

The density is needed for the modal analysis

Figure 8.13: Perform modal solution.

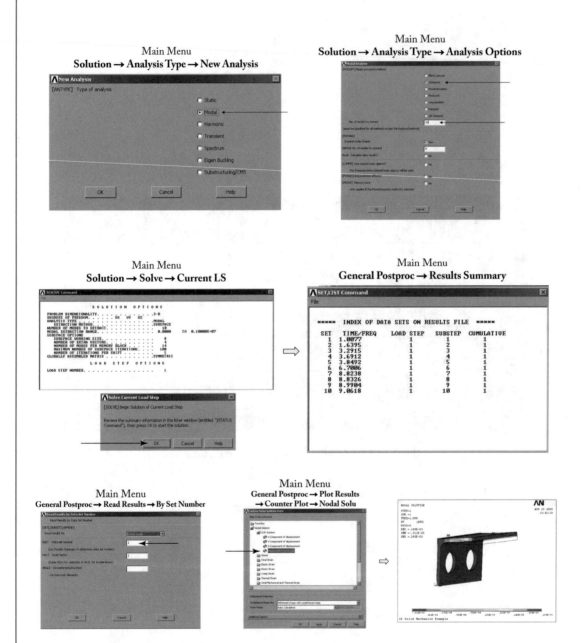

Figure 8.14: Perform modal solution.

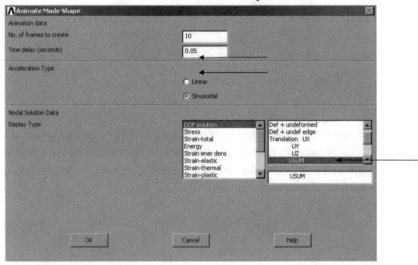

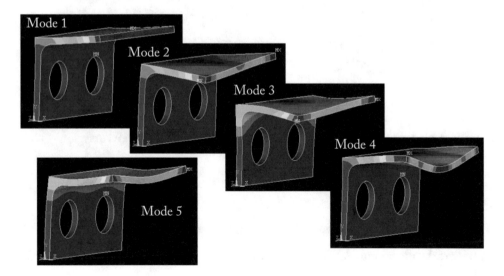

Figure 8.15: The mode shapes.

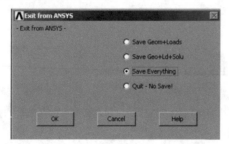

Figure 8.16: Exit from ANSYS.

Example 8.2   A cross-section of pipe fins with specific length created using Solidworks, is shown in Fig. 8.17. Use ANSYS to conduct heat transfer analysis to find the temperature distribution in the fins. The idea of this example is to illustrate the "parasolid" option in Solidworks to correctly transfer the model into ANSYS software.

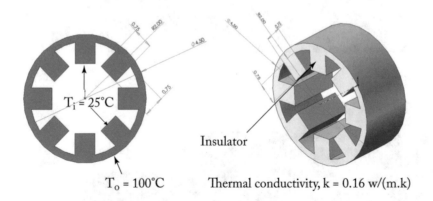

Figure 8.17: A pipe with fins for heat transfer analysis modeled using SolidWorks.

The following steps demonstrate how to create the model geometry and find a solution using ANSYS.

Step 1: Create the model in Solidworks and save the file as "IGES," as shown in Fig. 8.18.

Step 2: Import the model into ANSYS and check to make sure the model is transferred correctly, as shown in Fig. 8.19.

Step 3: It is evident that "IGES" file extension did not import the model into ANSYS correctly as there are no volumes in the model. So, let us try the other option of using "PARASOLID" file extension option in SolidWorks, as shown in Fig. 8.20.

- Create a solid model in SolidWorks

- Export as an IGES file
  – File > Save As
  – Save as type: Choose IGES
  – Click "Save"

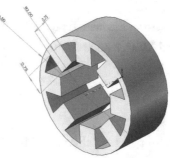

Figure 8.18: Create the model in Solidworks.

- File > Import > IGES
- Click "OK"

- Choose the IGES file

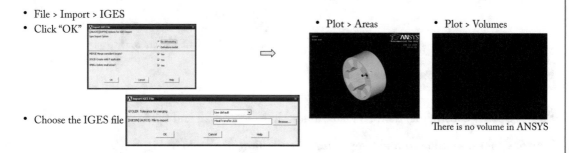

- Plot > Areas   - Plot > Volumes

There is no volume in ANSYS

Figure 8.19: Import the model into ANSYS.

- File > Import > PARA
- Choose the Parasolid file

- Plot > Volumes

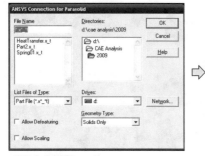

- Click "OK"

There is a volume in ANSYS

Figure 8.20: PARASOLID file.

Step 4: Conduct pre-processing by specifying the element type (BRICK 8 Node 70), material properties and meshing. See pre-processing activities (Fig. 8.21).

- Preprocessor > Element Type > Add/Edit/Delete
  - Add "Brick 8 node 70"

- Preprocessor > Material Props > Material Models

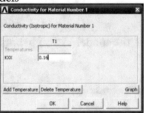

Figure 8.21: Conduct pre-processing.

Step 5: Specify constraints, thermal loads, as shown in Figs. 8.22 and 8.23.

Step 6: Perform solution to find temperature distribution, as shown in Fig. 8.24.

- Preprocessor > Loads > Define Loads > Apply >Thermal > Temperature > On Areas
  - Pick "outside area"
  - Type "373" for temperature

- Preprocessor > Loads > Define Loads > Apply >Thermal > Temperature > On Areas
  - Pick "inside area"
  - Type "298" for temperature

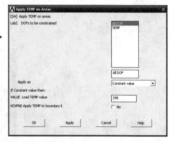

- Preprocessor > Loads > Define Loads > Apply >Thermal > Heat Flux > On Areas
  - Pick "front & back area"
  - Type "0" for heat flux

Figure 8.22: Constraints, thermal loads.

- Preprocessor > Meshing > Mesh Tool
  - Size Controls > Global > Set
  - Type "0.003" for size

  - Click "**Mesh**"

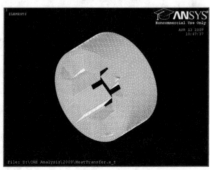

Figure 8.23: Constraints, thermal loads.

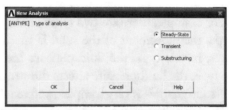

- Solution > Analysis Type > New Analysis
  – Choose "Steady-State"

- Solution > Solve > Current LS
  – Click "OK"

- General Postproc > Plot Results > Contour Plot > Nodal Solu
  – DOF Solution > Temperature

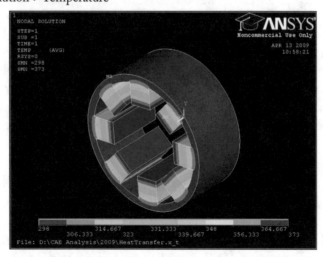

Figure 8.24: Find temperature distribution.

8.5   MULTI-PHYSICS PROBLEMS

Most practical engineering problems involve more than one discipline, either fluid or solid, and so on. Also, there is an increasing demand to solve realistic problems in a variety of engineering disciplines which require using multiphysics simulations. These include, fluid-structure systems, thermal-fluid, thermal-structure, and structure-electro-magnetic systems.

For example, there are two types of analysis related to fluid-structure multiphysics. In the first type, the movement of the solid is specified and the resulting fluid pressure is estimated. However, in this type, the fluid pressure has no effect on the movement of the solid. In the second type, the fluid pressure affects the structure deformation and vice versa. The second type of fluid-solid analysis is very complex, and requires intensive computational effort and experience.

In biomechanical analysis, say for example, lung airways, in which lung tissues are flexible, the second type of fluid-solid analysis needs to be carried out. In general, the computational procedure for multiphysics problems involves fluid and solid models. The solution to the fluid-structure interaction problem can be obtained by solving the governing equations for fluid and airway structure consecutively. At each time step, the algorithm begins by solving the flow equations to obtain fluid pressure. Airway structure equations are then solved for the displacement using the fluid pressure as an external force. The airflow equations are then solved again to obtain the fluid pressure after the airway displacement changes the fluid boundaries. This loop continues until both fluid pressure and airway displacement converge, as shown in Fig. 8.25.

Similarly, in thermal-fluid analysis, the fluid is solved first, and then using the results of fluid analysis as boundary conditions for the thermal analysis and the process is repeated until the solution converges. In ANSYS multi-physics package, there are various analyses which include ANSYS/Mechanical for structural and thermal analysis; ANSYS/Flotran for fluid dynamics analysis; ANSYS/Emag for magnetic field analysis; and ANSYS/LS-DYNA for dynamic analysis. The following multi-physics example illustrates the various steps involved in the solution to a fluid-solid analysis. For additional example problems, the reader is referred to the ANSYS manual.

8.6   MULTI-PHYSICS (FLUID-SOLID INTERACTION) EXAMPLE

Example 8.3   In this example, you will learn how to solve a multiphysics problem by using Fluid-Structure Interaction (FSI) simulation in ANSYS Workbench and employ both ANSYS Mechanical and ANSYS Fluent solvers at the same time. The objective of this simulation is to estimate airflow characteristics such as velocity, pressure, shear stress, and mechanical properties such as strain and Von-mises stress on human lung bifurcation through fluid and solid interaction analysis.

A 3D human lung bifurcation model is designed, as shown in Fig. 8.26 based on Magnetic Resonance Image (MRI) data using AutoCad and Solidworks software. In this model, air flows

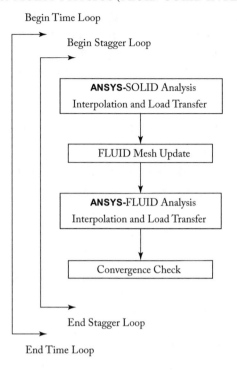

Figure 8.25: Fluid-solid multiphysics algorithm in ANSYS.

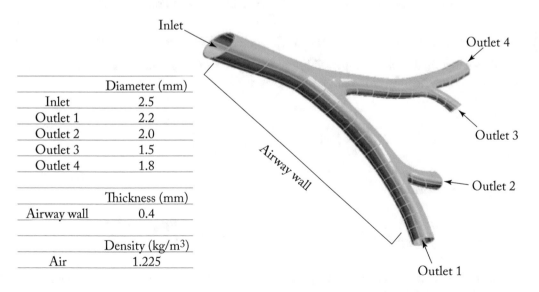

	Diameter (mm)
Inlet	2.5
Outlet 1	2.2
Outlet 2	2.0
Outlet 3	1.5
Outlet 4	1.8

	Thickness (mm)
Airway wall	0.4

	Density (kg/m^3)
Air	1.225

Figure 8.26: Geometry and dimensions of the model for fluid-solid multi-physics analysis.

through the lung bifurcation from the inlet and exits from the four outlets. The diameter of the bifurcation at the inlet is approximately 2.5 mm. The diameter of outlet 1 is approximately 2.2 mm, the diameter of outlet 2 is around 2.0 mm, the diameter of outlet 3 is around 1.5 mm, and the diameter of outlet 4 is around 1.8 mm. The wall thickness is 0.4 mm. The density of air is 1.225 kg/m^3.

8.6.1   GOVERNING EQUATIONS DEFINING THE PHYSICS OF AIRWAY WALL

The airway wall is assumed to be an isotropic elastic linear material in the structural domain. The material properties developed in the airway are modeled using the Cauchy momentum relationship:

$$\frac{\partial \sigma_{ij}}{\partial x_j} + F_i = \rho \frac{\partial^2 u_j}{\partial t^2}.$$

The constitutive equation of a linear elastic material is derived as follows:

$$\sigma_{ij} = C_{ijkl}\varepsilon_{kl},$$

where σ is the stress in each direction, F is the body force, ρ is mass density, u is the displacement, C is the elasticity tensor, and $varepsilon$ is the strain in each direction.

Material properties employed in this problem: the density for airway wall is 1365.6 kg/m^3 and the Young's modulus (elastic modulus) is 75 kPa, and Poisson's ratio is 0.45.

8.6.2   GOVERNING EQUATIONS DEFINING THE PHYSICS OF FLUID

The airflow is assumed to be incompressible (air density is constant), and the mass and momentum of airflow through the bifurcation airway are governed by Navier-Stokes equations.

Conservation of mass

$$\frac{\rho}{\sqrt{g}}\frac{\partial}{\partial t}\left(\sqrt{g}\right) + \rho\frac{\partial}{\partial x_j}\left(u_j - \frac{\partial \tilde{x}_j}{\partial t}\right) = 0.$$

Conservation of momentum

$$\frac{\rho}{\sqrt{g}}\frac{\partial}{\partial t}\left(\sqrt{g}u_i\right) + \rho\frac{\partial}{\partial x_j}\left[\left(u_j - \frac{\partial \tilde{x}_j}{\partial t}\right)u_i\right] = -\frac{\partial p}{\partial x_i} + \mu\frac{\partial^2 u_i}{\partial x_j^2},$$

where $\tilde{x}$ represents the moving mesh location, $\sqrt{g}$ is the metric tensor determinate of the transformation, that is, the local computational control-volume size, ρ is fluid density, p is fluid pressure, μ is fluid viscosity, and u is fluid velocity.

8.6.3   INPUT AND BOUNDARY CONDITIONS

A flow rate boundary condition is imposed at the inlet and no-slip boundary conditions are assumed at the bifurcation airway walls in this problem. A pressure outlet boundary condition at the four outlets is assumed to be a constant (0 Pa) for this problem.

Mechanical ventilation (transient—unsteady) breathing waveform, which is a time-varying boundary condition, is imposed at the inlet. Mechanical ventilation waveform is developed as a user-defined function (UDF) of airway surface area and lung generation number. The equation for velocity with varying time in mechanical ventilation waveform is derived as follows:

$$u(t) = Q/F(A) \times 2^{n-1} \qquad \text{(Inhalation)}$$
$$u(t) = -Q \times \exp(t/0.4)/F(A) \times 2^{n-1} \qquad (Exhalation),$$

where Q is flow rate (tidal volume/inhalation time), $F(A)$ is total surface area of bifurcation, and n is generation number. Tidal volume is assumed to 420 mL:

- to start ANSYS Workbench on Windows, click the Start menu, then select All programs > ANSYS 16.0 > Workbench 16.0.

The project schematic appears as an Unsaved Project, as shown in Fig. 8.3. 0.4 s–2 s, as shown in Fig. 8.27.

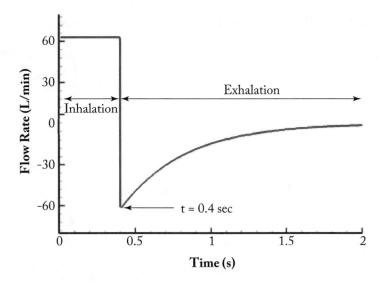

Figure 8.27: Mechanical ventilation breathing waveform imposed at the inlet as boundary condition for fluid.

Solution: The following illustrates a step-by-step process for finding the solution.

Step 1: Start ANSYS Workbench.

- To start ANSYS Workbench on Windows, click the Start menu, then select All programs > ANSYS 16.0 > Workbench 16.0.

- Create a directory where you will store your project (this is your working directory).

- Select File > Save or click Save.

- Select the path to your working directory to store files during this tutorial.

- Under File name, type LungBifurcation (you can also choose any other file name) and click Save.

- To make the Files view visible, select View > Files from the main menu of ANSYS Workbench.

   The project schematic appears as an Unsaved Project, as shown in Fig. 8.28.

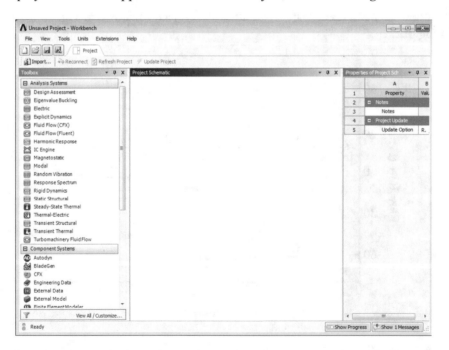

Figure 8.28: Unsaved project.

Step 2: Import geometry model to ANSYS Workbench through an existing DesignModeler file.

- From the Component Systems toolbox located on the left side of the ANSYS Workbench window, select the Geometry template. Double-click the template, or drag it onto the Project Schematic. Now, the project schematic should be displayed, as shown in Fig. 8.29.

- On the project Schematic, right-click the Geometry's Geometry cell (A2) and select import Geometry > Browse.

- In the open dialog box, browse your working directory, select the file you run from your working directory, and click open.

- In the Geometry system of the project schematic, double-click the Geometry cell (A2) to edit the geometry using DesignModeler. The DesignModeler application opens in a separate window.

- The geometry is set up for the project. Save any changes by selecting File > Save > Project from the main menu in DesignModeler, and then select File > Close DesignModeler to return to the Project Schematic.

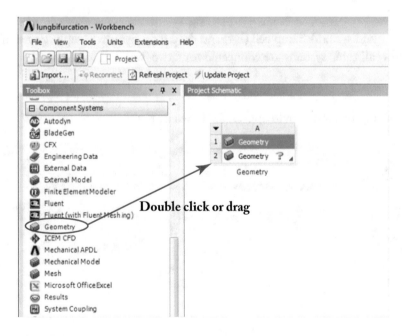

Figure 8.29: Project schematic.

Step 3: Adding Analysis Systems to the Project: FSI analysis is conducted by adding three analysis systems: a Transient Structural system and a Fluid Flow (Fluent) system. Then, use the System

Coupling to couple the other two systems and to coordinate the solution execution. The three systems are added to the Workbench project as follows.

- From the Analysis Systems toolbox located on the left side of the ANSYS Workbench window, select the Transient Structural template. Double-click the template, or drag it onto the Project Schematic to create a Transient Structural.

- Similarly, drag a Fluid Flow (Fluent) in analysis system on the toolbox in the left of the project schematic.

- Expand the Component Systems on toolbox, drag a System Coupling system and drop it to the right of the Fluid Flow (Fluent) system.

- Drag the Geometry cell (A2) and drop it on the Transient Structural geometry cell (B3), and also drag the Geometry cell (A2) and drop it on the Fluid Flow system geometry cell (C2).

- Drag the Structural system's Setup cell (A5) and drop it on the System Coupling system's Setup cell (C2).

- Drag the Fluid system's Setup cell (B4) and drop it on System Coupling system's Setup cell (C2). Now all three systems are connected for a two-way FSI analysis.

- Save the project.

The Project Schematic should appear, as shown in Fig. 8.30.

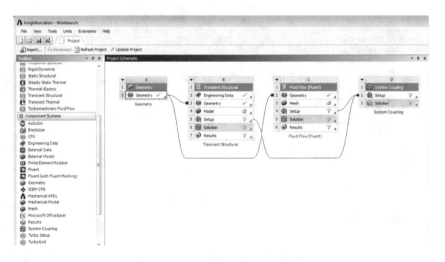

Figure 8.30: Adding analysis systems. Project schematic.

Now, the geometry will be shared between the Structural and Fluid systems with the same geometry because they are connected by their geometry cells in the Project Schematic. Now the Set up in Transient system and Set up in Fluid Flow are connected to Set up in System coupling.

Step 4: Adding a New Material for the Project. In the Project Schematic, the Structural system's Engineering Data cell (B2) appears in an up-to-date state because default material is already available for the project. You will use this material for the lung bifurcation, and so you need to update this cell by adding this new material to the Engineering Data as follows.

- On the Project Schematic, double-click the Engineering Data cell (A2) in the Structural system. Engineering Data opens in a new tab in Workbench, as shown in Fig. 8.31.

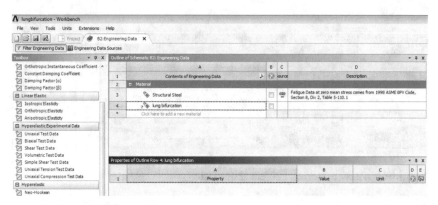

Figure 8.31: Engineering data.

- In the Outline of Schematic B2: Engineering Data view, click the empty row at the bottom of the table to add new material for the project such as lung bifurcation (Fig. 8.31).

When you click away from that cell, lung bifurcation is created and appears with a blue question mark (Fig. 8.31), indicating that its properties need to be defined.

From the Toolbox on the left, expand Physical Properties. Select Density and drag it onto the cell containing lung bifurcation (A4) in the Outline of Schematic A2:

Density is added in the Properties of Outline Row 4.

- In the Properties of Outline Row 4: set the Value of Density (B2) to 1365.6 kg/m^{-3}. Do not type in units.

- In the toolbox under Linear Elastic, drag Isotropic Elasticity onto bifurcation (A4) in the Outline of Schematic A2.

Isotropic Elasticity is added as the lung bifurcation property in the Properties of Outline Row 4.

- In the Properties of Outline Row 4: expand Isotropic Elasticity by clicking the plus sign. Now set Young's Modulus to 75,000 [Pa] and Poisson's Ratio to 0.45. Do not type in units.

Now, your Project Schematic appears, as shown in Fig. 8.32.

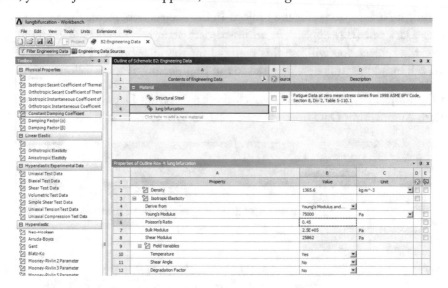

Figure 8.32: Adding a new material. Project schematic.

- From the main menu, select File > Save to save material settings for the project.

- Close the Engineering Data tab to return to the Project Schematic.

Step 5: Defining the Physics in the Mechanical Application

In the Mechanical application, you are setting up the structural analysis and defining the coupling interface (boundary condition). You will not solve the structural analysis from the Mechanical application because you will use the System Coupling system to solve both structural and fluid systems at the same time. The structural Geometry cell (B3) is up-to-date, and so you start your setup by generating the structural mesh. This section describes the step-by-step definition of the structural physics.

Generating the Mesh for the Structural System

Generate the mesh for the Structural system directly in the Mechanical application.

- On the Project Schematic, double-click the Structural system's Model cell (B4) to open the Mechanical application. The Mechanical application opens in a separate window.

- In the Mechanical application's Outline on the left, expand Geometry to see the two geometries, shell (solid) and fluid.

- For the structural analysis, you need to generate the mesh for only the solid body. To do this, you need to first suppress the fluid body. Right-click the fluid geometry (which contains all of the fluid body), and select Suppress Body.

The fluid bodies are now suppressed and their status changes to an x mark. You now will see only the solid body in the Graphics view.

You will define the mesh by marking divisions on the edges of the solid. These divisions will be used as guides for the mesh creation.

- Click on Mesh in Outline.

Now, you see the Details of "Mesh," as shown in Fig. 8.33.

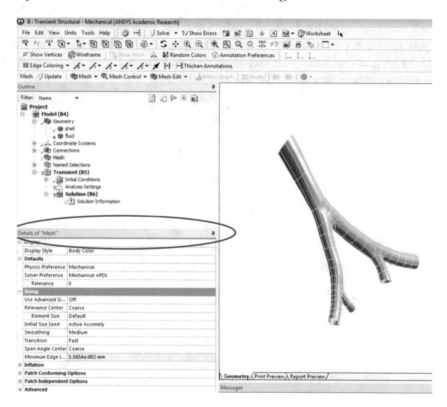

Figure 8.33: Details of Mesh.

- Change coarse to fine in Relevance Center

- Type element size to 0.1. If you need finer mesh, then reduce the size of the element.

In the Outline, right-click Mesh and select Generate Mesh.

Step 6: Assigning the Material to the Geometry

When you defined the lung bifurcation material, you set it to be the default for your solid body. In the Mechanical application, you can see that this material is set correctly.

- In the Mechanical application's Outline on the left, select Project > Model > Geometry > shell.

- In the Details of "shell," ensure that Material > Assignment is set to lung bifurcation.

Setting the Basic Analysis Values

You now need to set up information about the transient analysis' time steps, which are the basic analysis values needed for the transient structural analysis. The time step (0.01 s) is chosen to be an appropriate size to observe the lung bifurcation and the time duration (2 s) is also selected. These time settings are dependent on the physics that you are observing, including the material properties of the lung bifurcation. When setting your own transient analysis, make sure that you choose time settings appropriate to the physics you are solving.

- In the Mechanical application's Outline view, select Project > Model > Transient > Analysis Settings.

    The details of Analysis Settings appear in the Details of "Analysis Settings" below the Outline view, as shown in Fig. 8.34.

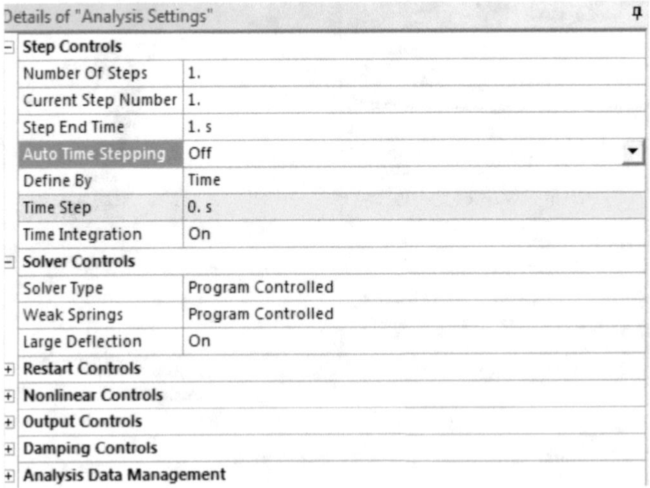

Details of "Analysis Settings"		무
Step Controls		
Number Of Steps	1.	
Current Step Number	1.	
Step End Time	1. s	
Auto Time Stepping	Off	▼
Define By	Time	
Time Step	0. s	
Time Integration	On	
Solver Controls		
Solver Type	Program Controlled	
Weak Springs	Program Controlled	
Large Deflection	On	
Restart Controls		
Nonlinear Controls		
Output Controls		
Damping Controls		
Analysis Data Management		

Figure 8.34: Analysis settings.

In the Details of "Analysis Settings," specify the following settings under Step Controls (do not type units next to the time values).

- Set Step End Time to 2.

- Set Auto Time Stepping to Off.

- Set Time Step to 0.01.

Defining the Fixed Support

The fixed support is needed to hold the inlet of lung bifurcation. Set up the fixed support.

- Right-click Transient in the Outline view, and select Insert > Fixed Support

- Select the face near the inlet, then the face is highlighted, as shown in Fig. 8.35.

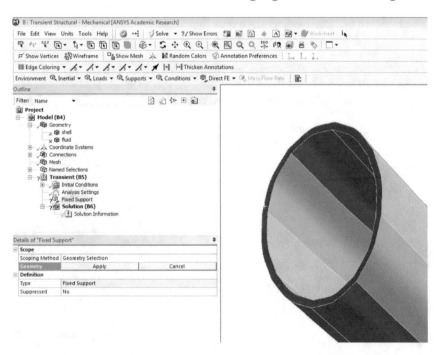

Figure 8.35: Face.

- In the Details of "Fixed Support" view, click Apply under? Geometry to set the fixed support.

Defining the Fluid-solid Interface

The fluid-solid interface defines the interface between the fluid in the Fluid system and the solid in the Structural system. Data will be exchanged across this interface during the execution of the simulation. When setting up your structural system for a coupled analysis, you need to define this interface on regions in the structural model that will receive force data from the Fluid system.

- In the Outline view, right-click Transient and select Insert > Fluid Solid Interface.

- Change Geometry selection in Named selection in Scoping Method of Details of Fluid Solid Interface.

- Using name selection in Details of Fluid Solid Interface, select the wall_cfd_coupled.

The interface will be displayed, as shown in Fig. 8.36.

Figure 8.36: The interface.

Completing the Setup for the Structural System

On the Project Schematic, the Structural system's Setup cell (B5) appears in an update-required state. To complete the setup in the Structural system, you need to ensure that all the data is in the right state in the Project Schematic.

- In the Structural system, right-click the Setup cell (B5) and select Update.

- From the main menu, select File > Save to save the project.

The set up for the Structural system is complete. Remember that you will not solve the structural analysis from the Mechanical application because you are using the System Coupling system to solve both Structural and Fluid systems at the same time. In the next section, you will set up the Fluid system.

Step 7: Setting up for Fluid Analysis

You will use the Fluent application to set up your Fluid system, but first you need to generate the mesh using the Meshing application. The fluid Geometry cell (C2) is up-to-date because it shares the geometry with the structural analysis, and so you start your Fluid system's setup with creating a mesh.

Generating the Mesh for the Fluid System

You will generate a mesh for the Fluid system using the Meshing application.

- In the Project Schematic, double-click the Fluid system's Mesh cell (C3) to open the Meshing application. The Meshing application appears in a separate window.

- In the Meshing application's Outline view on the left, expand Geometry to see the two geometries, shell and fluid.

- For the fluid analysis, you need to generate the mesh for only the fluid body. To do this, you need to first suppress the shell body.

  Right-click solid and select Suppress Body.

  The shell body is now suppressed and its status changes to an x mark. You will now only see the fluid body.

- In the Outline on the left, click Mesh. In the Details of "Mesh" below, under Defaults, notice that the Physics Preference is set to CFD and Solver Preference is set to Fluent.

- Now click the plus sign on Sizing in the Details of "Mesh."

- Change Coarse in Fine in Relevance Center

- Now that all of the settings for your mesh are complete, you need to generate the mesh. In the Outline, right-click Mesh and select Update.

- Select File > Save Project, and then File > Close Meshing to close the Meshing application.

Defining the Physics in the ANSYS Fluent Application

In the Fluent application, you are setting up the fluid analysis, and defining the coupling interface. You will not solve the fluid analysis from the Fluent application because you are using the System Coupling system to solve both structural and fluid systems at the same time.

Adding the Solution Setup Settings

You now need to open your analysis in the Fluent application, set the Fluid analysis to be transient, and add material to the fluid geometry.

- In the Project Schematic, double-click the Fluid system's Setup cell (C4) to open the Fluent application.

- The Fluent Launcher opens in a new window. Under Options, select Double Precision.

- Use the remaining default options (3D and serial), and click OK to close the Fluent Launcher.

  The Fluent application opens in a new window, and the mesh file is automatically loaded.

- On the left, select Solution Setup > General. Under Time, click the Transient option.

- On the left, select Solution Setup > Materials > Air to assign material to your geometry. Click the Create/Edit button, and in the dialog box that appears, Use default value for Density and Viscosity of air.

  Click Change/Create to save these changes, and then click Close.

- Under Solution Setup > Models, note that by default, the viscous model is laminar and the energy model is turned off. No changes are needed to these settings.

Adding the Mechanical Ventilation waveform at Inlet

The air entering the inlet is inhaled in mechanical ventilation waveform, so you need to import waveform at inlet. To import the User Define Function file for the waveform follow the steps below:

- Define > Functions > complied, then you will see Complied UDF's window.

- Click Add, and find the UDF file in your directory, then click build > click Load.

Now that your waveform is imported, you need to apply this waveform at the inlet.

- Select Solution Setup > Boundary Conditions > inlet > Click Edit > Click arrow on constant of Velocity Inlet > Change in UDF unsteady velocity > Click OK.

Now the mechanical ventilation waveform is applied at your inlet.

Defining the Dynamic Mesh

A dynamic mesh is needed for any coupled analysis where a system receives displacements. In this example, the wall of lung bifurcation is moving back and forth, and the dynamic meshing settings determine how the mesh of the fluid body reacts to this deformation of the moving shell (structural) body. The mesh on the fluid-structural interface is static, so as the fluid mesh is modified to accommodate the deformation in the transient system, and the mapping on this coupling interface stays consistent.

Set up the dynamic mesh:

- On the left, select Solution Setup > Dynamic Mesh.

- Check the Dynamic Mesh option in the panel.

The settings for Dynamic Mesh are now available.

- Under Mesh Methods, Smoothing is checked by default. Also, check the Remeshing box. Click the Settings button to specify the settings for the smoothing and remeshing used.

The Mesh Method Settings dialog box appears.

- On the Smoothing tab, set Method to Diffusion.

- For the Diffusion Parameter, type 2. Click OK to close the Mesh Method Settings dialog box.

For remeshing, leave as default.

- Under Dynamic Mesh Zones, click Create/Edit to specify which zones in your geometry will have dynamic meshing.

The Dynamic Mesh Zones dialog box appears. Define the dynamic mesh settings needed for the surface "wall_fea_couple," which is the wall of fluid. This surface will be affected by the solid body's displacement, and its mesh needs to be able to deform.

- In the Dynamic Mesh Zones dialog box, under the Zone Names drop down list, select the zone "wall fea couple."

- Set its Type as System coupling. Then click Create

- Similarly, repeat to select inlet, outlet 1, 2, 4, and 5, respectively, and Set its Type as Deforming respectively, then click Create.

Now, the list of Dynamic Mesh Zones now includes the "inlet," "outlet1," "outlet 2," "outlet 4," "outlet 5," and "wall fea coupled," as shown in Fig. 8.37.

Figure 8.37: Dynamic mesh zones.

You now have six dynamic mesh zones defined and listed on the right of the dialog box. Click Close.

Adding the Solution Settings

Set the solutions settings in the Fluent application so that your fluid system is ready to be solved.

- On the left side of the Fluent application, select Solution > Solution Methods.

    - Under Pressure-Velocity Coupling > Scheme, select Coupled.
    - Under Spatial Discretization > Momentum, select Second Order Upwind.

- On the left side of the Fluent application, select Solution > Calculation Activities, then specify Autosave Every (Time Steps) to be 10.

- On the left side of the Fluent application, select Solution > Run Calculation, then complete the following steps.

  - Specify Number of Time Steps to be 200. Note that the system coupling's number of time steps will override this value.
  - Specify the Max Iterations/Time Step to be 20. This value is the maximum amount of times that Fluent can iterate within a coupling iteration.
  - Specify the Time Step Size (s) as 0.01.

- On the left side of the Fluent application, select Solution > Solution Initialization. Under Initialization Methods, click the Standard Initialization option. Then select inlet from "Compute from."

- In Solution > Solution Initialization, click Initialize.

- Save the project.

- Select File > Close Fluent to close Fluent and to return to the Project Schematic.

The setup for the Fluid system is complete. Remember that you will not solve the fluid analysis from the Fluent application because you are using the System Coupling system to solve both structural and fluid systems at the same time. In the next section, you will set up the System Coupling system.

Step 8: Defining and Running the Coupling in the System Coupling Application

In the System Coupling system, you are setting up the coupling between your Structural and Fluid analyses. You will use the System Coupling system to solve both of these analyses at the same time.

Notice that in the Structural and Fluid systems, all of the cells up to Setup are marked as up-to-date.

Setting the Basic Analysis Values

To set up the transient analysis settings for your coupled analysis follow the steps below:

- In the Project Schematic, double-click the System Coupling system's Setup cell (D2).

In the dialog box, click Yes to allow upstream data to be read. The System Coupling system is obtaining data from the Structural and Fluid systems' Setup cells (B5 and C4). The System Coupling application opens in a new tab in your Workbench project.

- In the Outline of Schematic: System Coupling, select System Coupling > Setup > Analysis Settings.

- In Properties of Analysis Settings (on the bottom left):

  Set Analysis Type to Transient.

- In Properties of Analysis Settings (on the bottom left):

  Set Duration Controls > End Time to 2.

- The end time is the same as the Structural system's time duration. System Coupling's end time value always overrides the number of time steps specified in the Fluent application.

  Set Step Controls > Step Size to 0.01.

The coupling iteration size is the same as the transient analysis' time step, and the choice of 0.01 s is small enough for use to observe the bifurcation moving. System Coupling's step size value always overrides the time steps size specified in the Fluent application.

Creating the Data Transfers

For your coupled analysis, data from the Structural and Fluid solutions need to be shared throughout the solution process. System Coupling coordinates the transfer of data between these two systems using the Data Transfers that you create.

- In Outline of Schematic: System Coupling, expand System Coupling > Setup > Participants until all region components are visible.

- Ctrl-select the "wall_fea_coupled" (from the Fluid system) and "Fluid Solid Interface" regions (from the Structural system). With both selected, right-click on one of those regions and select Create Data Transfer.

- Under System Coupling > Setup > Data Transfers, Data Transfer and Data Transfer 2 are created:

  - Data Transfer: here, the surface of the Fluid system around the bifurcation transfers force to the surface of the Structural system around the bifurcation.

  - Data Transfer 2: here, the surface of the Structural system around the bifurcation transfers displacement to the surface of the Fluid system around the bifurcation.

Preparing System Coupling for Restarts

You should ensure that System Coupling is producing restart data, in the event that the System Coupling analysis needs to be restarted.

- Under System Coupling > Setup > Execution Control, select Intermediate Restart Data Output. The restart output frequency for the system coupling analysis is defined and controlled by these settings.

- In Properties of Intermediate Restart Data Output:

    - Set Output Frequency to at Step Interval.
    - Set Step Interval to 5.

- Select File > Save to save your settings before solving.

Solving and Restarting the Coupled Analysis

During the solution process, the System Coupling system coordinates the solving of your Structural and Fluid systems as well as the data transfers between these two systems. The Fluid system solves using the Structural solution's displacement data, and the Structural system solves using the Fluid solution's force data.

- To start solving the coupled analysis, in Outline of Schematic C1: System Coupling, right-click Solution and select Update.

- The System Coupling solution is complete when the System Information view reads "System coupling run completed successfully."

- Select File > Save to save the project, and then click on the Project tab to return to the Project Schematic.

Step 9: Viewing Results in CFD-Post

You will use CFD-Post to view the results of your coupled analysis. You have simulated the lung bifurcation with inhaled air at inlet. The results you have obtained show pressure, shear stress and strain for lung bifurcation under mechanical ventilation breathing cycle. You will be able to use CFD-Post to see the contours.

In Workbench, you need to set up the Project Schematic so that CFD-Post can read the solution of your Structural and Fluid systems.

To view the results in CFD-Post:

- In the Project Schematic, drag the Structural Solution cell (B6) to the Fluid Results cell (C6).

- Double-click the Fluent Results cell (C6) in the Fluid system to launch CFD-Post.

CFD-Post opens in a new window. Both sets of results are loaded into the CFD-Post session, and are ready for you to view.

Results on the Shell (Solid)

You will use a contour to display the strain of the solid body.

- Click insert tab > Contour > Change name as strain > OK

- Click □ of Variable in Details of strain > click strain > click Elastic Equivalent Strain > Click OK > Click Apply

    The strain contour is shown as displayed in Fig. 8.38.

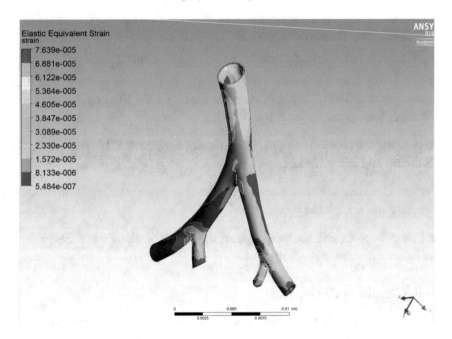

Figure 8.38: The strain contour.

Results on the Fluid

You will use a contour to display the pressure (shear stress) of the fluid body.

- Click insert tab > Contour > Change name as pressure (shear stress) > OK

- Click

- Click Variable in Details of pressure (shear stress) > click strain > Click OK > Click Apply

    The pressure and shear stress contours should display as shown in Fig. 8.39.

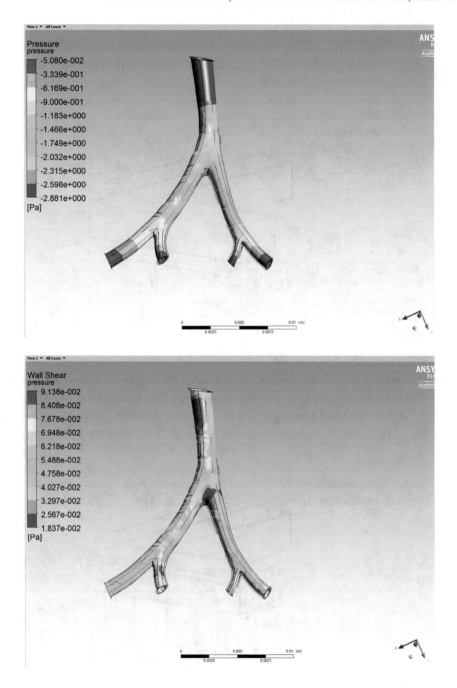

Figure 8.39: Pressure and shear stress contours.

8.7   EXERCISE PROBLEMS

1. The beam in Fig. 8.40 is used to support the indicated load. Using ANSYS, size the cross section of the beam. Standard 304 Stainless Steel is used with a modulus of elasticity of 190 GPa and yield strength 276 MPa.

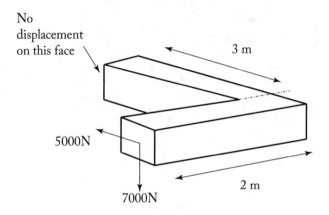

Figure 8.40: The beam.

2. The 304 stainless steel beam in Fig. 8.41 is used to support the indicated pressure of 7000 N/m² on the bottom face of the 2 m member. The material has a modulus of elasticity of 190 GPa and yield strength 276 MPa. Using ANSYS. Determine the maximum deflection of the member and the maximum stress. Will the support fail?

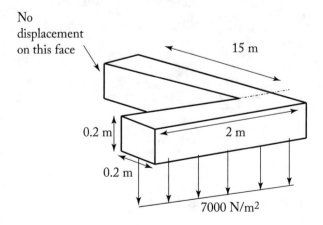

Figure 8.41: 304 stainless steel beam.

3. A ceramic mug is filled to the brim with hot coffee at 91°C and a thermal conductivity value of $k = 1.3$ W/m·k. The convection coefficient of inside the mug is $h = 50$ W/m²·k. Air at 23°C is blown over the entire mug resulting in a convection coefficient outside of $h = 150$ W/m²·k. Heat is also lost through the bottom of the mug which sits on a pine wood table at 20°C with a thermal conductivity value of $k = 0.11$ W/m·k, and a thickness of 2 cm. The table can be approximated as perfectly insulated. Using ANSYS, find the temperature distribution of the coffee mug and table (Fig. 8.42).

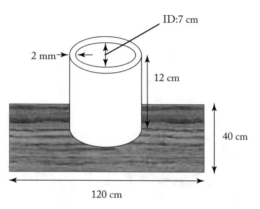

Figure 8.42: A ceramic mug.

4. A ceramic mug as shown in Fig. 8.42, is filled to the brim with ice cold whiskey at 0°C and a thermal conductivity value of $k = 1.3$ W/m·k. The convection coefficient inside the mug is $h = 20$ W/m²·k. Air at 23°C is blown over the entire mug resulting in a convection coefficient outside of $h = 150$ W/m²·k. Heat transfer also occurs through the bottom of the mug which sits on a pine wood table at 20°C with a thermal conductivity value of $k = 0.11$ W/m·k, and a thickness of 2 cm. The table can be approximated as perfectly insulated. Using ANSYS, find the temperature distribution of the coffee mug and the table.

5. Water enters a square pipe at a uniform velocity profile and is at 20°C with a viscosity of 1.002 mPa·s and a density of 998.2 kg/m³. The Pipe bends 90° after 5 m and another pipe splits off at 45° at 1 m after the bend. Find the velocity of the fluid at the exits of faces A and B using ANSYS (Fig.8.43).

6. Air enters a square pipe as shown in Fig. 8.43, at a uniform velocity profile at 28°C with a viscosity of 0.0186 mPa·s and a density of 1.172 kg/m³. The Pipe bends 90° after 5 m and another pipe splits off at 45° at 1 m after the bend. Find the velocity of the fluid at the exits of faces A and B using ANSYS.

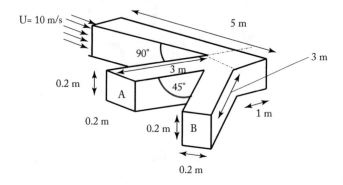

Figure 8.43: Square pipe.

CHAPTER 9

Sample Project

After reading this chapter, you will be able to:

- better understand and implement engineering analysis and simulations;
- carry out the analysis using ANSYS; and
- appreciate the capabilities of FE simulations.

9.1 OVERVIEW

This chapter introduces a sample design project to provide the reader with additional real life examples of design analysis projects. The sample project will further illustrate how the systematic finite element analysis process is followed in order to achieve a better understanding of the design/product. Several engineering industries require students to demonstrate real project experience using computational simulations to test their engineering skills, and appreciate the team effort involved in engineering projects. One of the student projects from the author's course is discussed. A brief list of design projects previously assigned by the author is also presented.

9.2 BENEFITS IN CONDUCTING ANALYSIS THROUGH PROJECTS

Engineering design and analysis projects provide students with the following learning objectives and outcomes:

- applying the design concepts for analysis;
- applying creativity, imagination, and problem solving skills;
- working with fellow students on a common goal and learning from others; and
- finally, acquiring skills in real world design analysis.

9.3 SAMPLE PROJECT—ANALYSIS OF WING GEOMETRIES FOR MAV

One student group project from a previous design class conducted a project on natural wing designs for Micro-Air-Vehicles (MAV) applications. The details of geometric models and analysis results are presented below.

9.3.1   OBJECTIVE

Due to the military's needs for better surveillance, reconnaissance, and search and rescue missions, the development of Micro Air Vehicles (MAV's) is becoming very prominent. The inspiration for MAV's comes from nature, from the wing flapping motion of insects and animals. The goal of this project is to investigate three natural wing designs (bat, butterfly, and dragonfly wings) that can be used for MAV's applications by analyzing the designs for stress analysis.

9.3.2   ANALYSIS MODELS

All three natural wings were modeled using SolidWorks. Wing geometries, including the venation of the butterfly and dragonfly wings and the bone support structure of the bat wing, were modeled as closely as possible to their natural counterparts. All wing models had the same surface area of 403 mm^2 with a thickness of 0.25 mm and wire frame supports had the same width and thickness of 1 mm. To replicate the flexibility of the natural wings, low density polyethylene ($E = 200$–400 MPa; Yield strength $= 15$–20 MPa) material was used. A titanium alloy ($E = 1.04e11$ Pa; Yield strength $= 1030$ MPa) was used to model the support wire frames to provide a lightweight yet strong enough structure to withstand lift forces. The three different models geometries are presented in Fig. 9.1.

The finite element mesh models for the three wing geometries are presented in Fig. 9.2.

The boundary conditions and the loadings for the three wing geometries are presented in Fig. 9.3.

The analysis results of displacements and stresses for the three wing geometries are presented in Fig. 9.4.

9.3.3   DISCUSSION

The results of displacement and maximum stresses were important metrics in deciding which wing design was the least fragile and able to withstand the largest force. The better design is a wing that will have the lowest displacement and stress values. From the results presented in Table 9.1, it is evident that the butterfly wing is the best design as it showed more uniformity in distributing the force on its wing geometry. In summary, the analysis revealed that the butterfly wing design is a better design, mainly for the uniformity in distribution of stress in its support wire frame and for having the lowest deformation and the lowest maximum stress for a given force in comparison to other two wing designs. The three natural designs are different and serve different purposes. It is recommended that these designs be studied in more detail by incorporating more design features and redoing the analysis for better comparison.

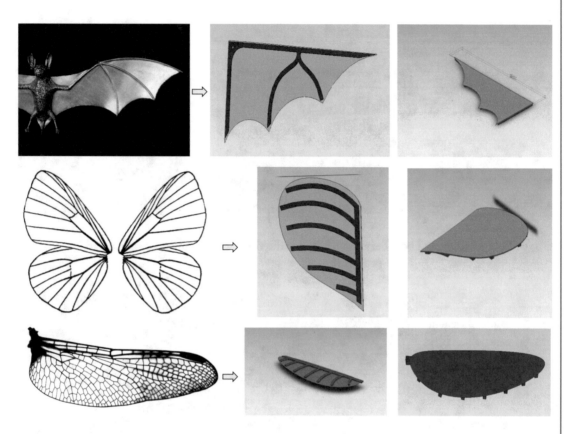

Figure 9.1: Three different wing geometries considered for analysis.

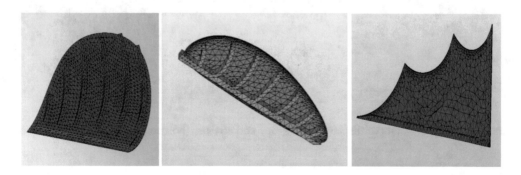

Figure 9.2: Finite element models created for the three wing geometries.

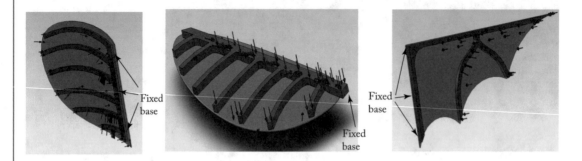

Figure 9.3: Loading and boundary conditions for the three wing geometries considered.

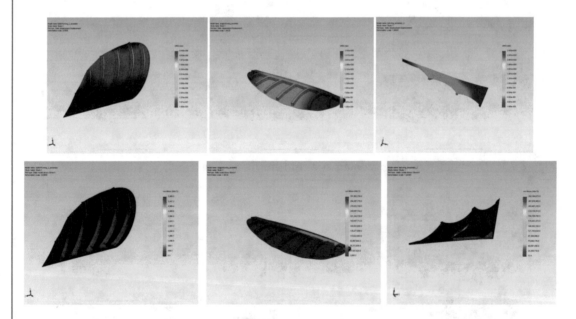

Figure 9.4: Analysis results of displacements (top) and stresses (bottom) for the three wing geometries.

Table 9.1: Analysis results for bat, butterfly, and dragonfly wing designs

Wing Design	Max Displacement (mm)	Max Stress (N/m^2)
Bat	2.4016	2.9220e+008
Butterfly	9.4457e-006	5986
Dragonfly	3.1818	3.3185e+8

9.4   EXAMPLE PROJECTS FOR ANALYSIS

The following is a brief list of design projects assigned by the author to his students.

- Design analysis of MEMS device.

- Analysis of power transmission lines.

- Analysis of a bridge in Richmond for stresses and vibration.

- Analysis of a baseball bat made from different materials.

- Analysis of bicycle frame made from different materials.

- Flow analysis through a flood wall.

- Heat transfer analysis of liquid in a cup.

- Structural analysis of a honeycomb design.

- Structural analysis of a natural bridge in Richmond.

- Fluid analysis in a lung bifurcation.

- Design analysis of NASA moonbuggy vehicle.

- Thermal analysis of a battery system design.

- Piezoelectric analysis of a sensor design.

- Fluid flow analysis of a drug delivery device.

Bibliography

[1] Alwadhi, E. M., *Finite Element Simulations Using ANYSYS*, 2nd Ed., CRC Press, 2015.

[2] ANSYS Theory and online manual, `http://www.ansys.com`

[3] Erdogan, M. and Ibrahim, G., *The Finite Element Method and Applications in Engineering using ANSYS*, 2nd Ed., Springer, 2015. DOI: 10.1007/978-1-4899-7550-8.

[4] Ellis, H. D., *The Finite Element Method for Mechanics of Solids with ANYSYS Applications*, Advances in Engineering Series, Taylor & Francis, 2011.

[5] Kim, N. and Sankar, B.V., *Introduction to Finite Element Analysis and Design*, John Wiley & Sons, Inc., 2009.

[6] Logan, D. L., *A First Course in Finite Element Method*, 6th Ed., Cengage Learning, 2016.

[7] Moaveni, S., *Finite Element Analysis-Theory and Applications with ANSYS*, Global Edition, Pearson Education Limited, 2015.

[8] Rao, S. S., *The Finite Element Method in Engineering*, 5th Ed., Elsevier Inc., 2011.

[9] Reddy, J. N., *The Finite Element Method in Engineering*, 5th Ed., Elsevier, Inc., 2011.

[10] Wael Al-Tabey, *Finite Element in Mechanical Design Using Ansys*, Lap Lambert Academic Publishing GmbH KG, 2012.

Author's Biography

RAMANA M. PIDAPARTI

Ramana M. Pidaparti earned his Ph.D. in aeronautics and astronautics from Purdue University, West Lafayette, Indiana. Dr. Pidaparti is now a professor and Associate Dean for Academic Programs in the College of Engineering at the University of Georgia. He was previously a faculty member at Virginia Commonwealth University and Purdue School of Engineering and Technology (IUPUI). He taught design and analysis courses emphasizing multidisciplinary designs and simulations, and encouraged students to participate in multidisciplinary design competitions. He won several design awards and presented papers at various conferences. He is a Fellow of ASME (American Society of Mechanical Engineers), Fellow of American Association for Advancement of Science (AAAS), Fellow of Royal Aeronautical Society, and Associate Fellow of AIAA (American Institute for Aeronautics and Astronautics). His current research interests include multidisciplinary design from Arts and Biomimicry, and bioinspired materials and structures, and computational mechanics and informatics.

Printed in the United States
by Baker & Taylor Publisher Services